NEUROLAW AND RESPONSIBILITY FOR ACTION

Law regulates human behaviour, a phenomenon about which neuroscience has much to say. Neuroscience can tell us whether a defendant suffers from a brain abnormality or injury, and it can correlate these neural deficits with criminal offending. Using fMRI and other technologies it might indicate whether a witness is telling lies or the truth. It can further propose neuro-interventions to 'change' the brains of offenders and so to reduce their propensity to offend. And, it can make suggestions about whether a defendant knows or merely suspects a prohibited state of affairs, so drawing distinctions among the mental states that are central to legal responsibility. Each of these matters has philosophical import; is a neurological 'deficit' inculpatory or exculpatory; what is the proper role for law if the mind is no more than the brain; is lying really a brain state, and can neuroscience really 'read' the brain? In this edited collection, leading contributors to the field provide new insights on these matters, bringing to light the great challenges that arise when disciplinary boundaries merge.

BEBHINN DONNELLY-LAZAROV is a professor at the Centre for Law and Philosophy, School of Law, University of Surrey. She is the author of *A Philosophy of Criminal Attempts* (2015) and *A Natural Law Approach to Normativity* (2007). Her recent and ongoing work aims to show that a proper grasp of mind and action is indispensable to a proper understanding of criminal responsibility.

NEUROLAW AND RESPONSIBILITY FOR ACTION

Concepts, Crimes, and Courts

Edited by

BEBHINN DONNELLY-LAZAROV

University of Surrey

WITH DENNIS PATTERSON AND PETER RAYNOR

CAMBRIDGE
UNIVERSITY PRESS

CAMBRIDGE
UNIVERSITY PRESS

University Printing House, Cambridge CB2 8BS, United Kingdom

One Liberty Plaza, 20th Floor, New York, NY 10006, USA

477 Williamstown Road, Port Melbourne, VIC 3207, Australia

314–321, 3rd Floor, Plot 3, Splendor Forum, Jasola District Centre, New Delhi – 110025, India

79 Anson Road, #06–04/06, Singapore 079906

Cambridge University Press is part of the University of Cambridge.

It furthers the University's mission by disseminating knowledge in the pursuit of education, learning, and research at the highest international levels of excellence.

www.cambridge.org
Information on this title: www.cambridge.org/9781108428705
DOI: 10.1017/9781108553339

First published 2018

Printed in the United Kingdom by Clays, St Ives plc

A catalogue record for this publication is available from the British Library.

Library of Congress Cataloging-in-Publication Data
Names: Donnelly-Lazarov, Bebhinn, editor. | Patterson, Dennis M. (Dennis Michael), 1955– editor. | Raynor, Peter, editor.
Title: Neurolaw and responsibility for action : concepts, crimes, and courts / edited by Bebhinn Donnelly-Lazarov, University of Surrey, with Dennis Patterson and Peter Raynor.
Description: Cambridge, United Kingdom ; New York, NY, USA : Cambridge University Press, 2018. | Includes bibliographical references and index.
Identifiers: LCCN 2017054438 | ISBN 9781108428705 (hardback) | ISBN 9781108450928 (pbk.)
Subjects: LCSH: Criminal liability–Psychological aspects. | Neurobiology–Social aspects.
Classification: LCC K5064 .N48 2018 | DDC 345/.04–dc23
LC record available at https://lccn.loc.gov/2017054438

ISBN 978-1-108-42870-5 Hardback

CONTENTS

FIGURES

CONTRIBUTORS

JOHN DANAHER is a lecturer in the Law School at NUI Galway. His research interests lie, broadly, in the areas of philosophy of law and emerging technologies and law. In the past, he has published articles on human enhancement, brain-based lie detection, the philosophy of punishment and artificial intelligence. He maintains a blog called Philosophical Disquisitions.

NICK J. DAVIS is a neuroscientist and a Senior Lecturer in Psychology at Manchester Metropolitan University. His work focuses on the use of brain stimulation technology to investigate and to enhance human cognition.

BEBHINN DONNELLY-LAZAROV is a Professor at the School of Law in at the University of Surrey. Her research considers the implications for law of the philosophies of mind and action. Bebhinn is a member of the Surrey Centre for Law and Philosophy.

JOANNA GLYNN QC is a Barrister, working in private practice until 2017. She acted as leading counsel in numerous homicide cases, especially infant deaths. Joanna is currently pursuing academic and research interests in the field of medical law at Queen Mary, University of London.

MARION GODMAN is a researcher at the Academy of Finland Centre of Excellence in Philosophy of the Social Sciences (TINT), Helsinki University, Finland. Her research interests span a range of topics in the philosophy of science, but Marion is particularly interested in ontological issues related to biology and the social sciences (e.g., kinds, scientific realism and unity of science).

PIM HASELAGER is a philosopher and psychologist, working at the Donders Institute for Brain, Cognition and Behaviour of the Radboud University, Nijmegen, the Netherlands. His work focuses on the ethical, legal and societal implications of Cognitive Neuroscience and Artificial Intelligence.

GIULIO MECACCI is a philosopher of technology, working at TU Delft and Radboud University. His research and teaching activity focuses on the ethical and societal implications of technology, ranging from neuroimaging to self-driving cars.

MICHAEL S. MOORE is one of most prominent authorities on the intersection of law and philosophy and widely regarded as the leading theoretician of the criminal law in the United States. Professor Moore joined the University of Illinois's College of Law in 2002 as the Charles R. Walgreen, Jr. Chair.

MICHAEL S. PARDO is the Henry Upson Sims Professor of Law at the University of Alabama School of Law. His scholarship focuses on the areas of legal evidence and proof, criminal procedure, civil procedure and philosophy of law.

DENNIS PATTERSON is a leading scholar in the field of Law and Neuroscience. He is the author or editor of four books in this emerging subfield of legal philosophy. He is the author of *Law and Truth* (Oxford University Press) and the editor of the *Oxford Introductions to U.S. Law*. He is also an expert in global trade law.

PETER RAYNOR is a former probation officer, now Emeritus Research Professor in Criminology and Criminal Justice at Swansea University. He has published widely on criminal justice topics over the last 40 years and is a Fellow of the Academy of Social Sciences.

ELIZABETH SHAW joined the law school at Aberdeen University, Scotland as a lecturer in September 2012. She undertook her PhD at Edinburgh University on the implications of free will scepticism for the criminal justice system. Her research interests are interdisciplinary, involving criminal law, philosophy and neuroscience.

KATRINA SIFFERD is Professor and Chair at Elmhurst College, Illinois, Department of Philosophy. Before becoming a philosopher, Katrina earned a Juris Doctorate and worked as a senior research analyst on criminal justice projects for the U.S. National Institute of Justice. Katrina is the author of numerous articles and book chapters on responsibility, criminal law, reductionism and punishment.

Introduction

Like other animals, we humans are distinctive. Physically, we have our own particular constitution. Behaviourally, we are special, acting, at least to an elevated degree, with purpose, sentience and self-consciousness. We are normative beings, having both an emotional and rational grasp of what we ought to do, framed by the existence of other conscious beings, by the environment we share, and by the institution of law. We experience great internal battles between our rational and appetitive powers and, depending on the outcome, may conform our behaviour, or not, to our normative commitments. We possess a range of interacting capacities – to move, to think, to comprehend, to want, to feel, to have reasons, and through these capacities, we are enabled to act. We are curious beings, pursuing knowledge not only for other ends but also for its own sake. And, although we are social and political, we are also intensely personal, existentially attuned to our condition in life. These depictions of human beings are depictions of a creature with a brain – with a human brain, of course. When it comes to understanding why we are the way we are perhaps there is nothing more to say: we are like *this* because our brains make us like this.

Each chapter in this multidisciplinary collection contributes, in some way, to understanding whether, how, and the extent to which this follows. More specifically, the chapters are concerned to examine the particularly prescient implications for legal responsibility of rapidly emerging neuroscientific understandings of the human brain. With unique authority, law regulates human action, a phenomenon about which neuroscience has much to say. By identifying neural correlates, it might tell us something that we did not previously understand about particular forms of culpable conduct; it can suggest interventions or neuro-enhancements to reduce an individual's propensity for criminal behaviour. It can illuminate the condition of the brain and speak of whether a defendant suffers a neurological 'deficit' or injury that might affect self-control, empathy, impulses, understanding or capacity for foresight. Using fMRI and other techniques, it may help a court to understand whether a witness is telling lies or the truth, and by observing impulses in the brain, it might speak to whether a defendant intends rather than merely suspects, say, a prohibited outcome. Insofar as it casts relevant light on an accused's state of mind, or on a witness's testimony and depending on court rules, it might constitute evidence, and so the need arises to understand the complex evidence itself and its legal status/import.

Such is its significance that the interaction between neuroscience and law is now the subject of a dedicated field of study, 'neurolaw' (a term that this collection happily appropriates). By necessity the field integrates a number of perspectives. Here, the contributions come from cognitive neuroscience, law, psychology, criminology and philosophy. From the varying perspectives, three broad themes, with particular importance for law, emerge: (1) whether brains are the locus of responsibility, including legal responsibility, and what it might mean to say that they are; (2) what kind of information neuroscience really does provide (to law, in particular); and (3) the relevance in the courtroom of what neuroscience has to say. The themes correspond to the book's sections that can usefully be characterised as, respectively; conceptual, epistemic and legal.

The first section of this book is about the conceptual relationship between brains and human behaviour. According to the various proposed accounts of that relationship, the implications for responsibility are then explored. Naturally reductionism occupies a central place in this discussion: can human behaviour and states of mind be reduced to the brain, and if so, what kind of reduction might this be? Does human action even correlate to brain states? Are some or all of these reductions fatal (or not) to responsibility? On these and related questions, Dennis Patterson and Michael Moore seem to occupy different points on a continuum rather than different planes entirely. Patterson suggests that we can do better than refer to brain states in explicating human behaviour. In this regard, he rejects materialism, identifying laws, baseball, and poetry as three actions in the world that are not material. He proposes that for reductionism to offer anything close to an explanation of actions like these, it would need to justify avoiding their contextual and intentional elements, a task that seems, to him, impossible. Patterson's alternative approach, rooted in teleology, is to say that an understanding of action must appeal to reasons, hermeneutics and contexts rather than to efficient causal and deterministic explanations.

For Moore, reductionism, whatever form it might properly take, nonetheless has its limitations. He proposes that we cannot reduce who we are to the individual neurons in the brain, if it is to follow from such a reduction that the self disappears. But, he shows that many ordinary reductions are actually quite harmless in this respect: even if these reductions suggest that human beings are ultimately machines, the idea survives them that we are wonderful machines with a very special mechanism indeed. An interesting aspect of Moore's position is the idea that actually the most meaningful challenge to our mentalistic explanations might be the wholehearted eliminative materialism (EM) that seeks to replace mind with brain, rather than to so reduce it. If true, Moore agrees with Fodor that EM is, an 'intellectual catastrophe'. Fortunately, Moore finds no reason to assume any truth in it.

Sifferd's chapter, like Moore's, offers us a forensic analysis of reductionist enterprises. Her aim is to assess the compatibility of these various reductionist accounts with the mental causation that criminal law depends upon. Which explanation of the relationship between brain and mind best supports law's account of criminal responsibility and the folk concepts it entails? Sifferd defends the view that noneliminative reductionism fits the bill. She proposes that a noneliminative reductivist may readily embrace the causal power of mental states. Specifically, the account can admit the possibility that folk concepts pick out a disjunction of local physical states and do so reliably enough for us to consistently use such concepts to predict and understand human behaviour. In any particular case, a mental state has causal properties as a particular token instantiation of this local disjunction. In this way, Sifferd's account offers support for law's mens rea/actus reus paradigm, and like Moore, she affirms that 'we' remain responsible for what we do notwithstanding the fact that our doings have neural correlates. Indeed, for her, noneliminative reductionism supports rather than challenges or defeats this insight.

Donnelly-Lazarov considers the concept of intention, making the case that human intentions cannot be reduced, in any way, to brain states. Taking up a recent trend in the philosophy of action, her suggestion is that we know our actions non-observationally and that that this, alone, is what it means to intend. Moreover, by her account, there is no sensation of intending that accompanies a sensation of acting; the one is the other. For Donnelly-Lazarov, it follows that intention is not the kind of thing that human beings can be said to have at all, in the mind or the brain or anywhere else. She suggests then that neuroscience will have to overcome a conceptual minefield if it is to account for intention in physical terms.

Nick Davis considers Patterson's view that reasons provide a better explanation for behaviour than physicalist explanations. He partly agrees but only for the time being. For Davis, law and neuroscience are natural friends. Legal scholars, he claims, will find it profitable to understand the basis of voluntary action and its underpinnings in the brain. However, he cautions that lawyers must also understand the limits of current understanding about the functions of the brain, both in terms of the technological limits of the machines we use to probe the brain and also in terms of the questions we are able to ask of the brain. Although, to this extent, he agrees with Patterson that current neuroscientific explanations of action are unsatisfying, he urges the neuroscientists to keep trying.

The second theme that the book explores is an epistemic one; it is about the nature and proper use of the information that our brains can provide. To begin, Mike Pardo examines the claims of lie-detection experiments. He notes that although the empirical problems associated with these experiments have received much attention, there are conceptual issues that warrant further scrutiny. Of particular concern to Pardo are the (false) presuppositions, embedded in these experiments, about what exactly is being measured. Pardo emphasizes the point that in order to confirm that brain activity correlates with any variable *X*, correct conceptual accounts of what exactly constitutes *X* are required. In particular, in the context of fMRI lie detection, scientists need to know what 'lying' is before they can identify its neural correlates. Pardo proceeds to analyse the (poorly understood) distinction between lying and deception and to scrutinise the concept of lying employed in the studies.

John Danaher also looks at the specific area of lie detection. He is less sceptical than some about its merits. In particular, Danaher suggests that EEG-based lie-detection tests, used appropriately, can assist the courts. Perhaps more important, these techniques, he claims, withstand philosophical opposition. Danaher includes, among this apparently ineffective opposition, the neurolaw mereological fallacy, identified by Patterson and Pardo (wherein states of mind or actions are wrongly attributed to the brain rather than to the whole person).

Forming the backdrop to both Danaher's and Pardo's accounts is some scepticism about the idea that neuroscience allows us to read the brain (or at least about whether it yet so allows). Does it? Searle has noted that computers do not really defeat humans at playing chess because the machines are not really *playing* chess at all.[1] It might similarly be suggested that *reading* the brain is not really what neuroscience does. This is precisely Haselager and Mecacci's charge. The authors ask of fMRI techniques, for example, whether they 'understand' content meaning and suggest that contentwise such techniques are opaque. Haselager and Mecacci use the analogy of a cash point machine 'reading' a pin number. What is the pin machine doing with the code? They claim that a communication with the pin machine is not one of content; the machine receives an opaque instruction that allows it to identify the correct vehicle but not to *understand* the inputted content. This is the same limitation of brain reading as Haselager and Mecacci see it: The current stage of neuroscience does not enable it to read content. Rather, it allows it to categorise patterns of vehicle similarity. In essence, brain reading gives no systematic access to propositional attitudes. Why might this matter to law? The authors touch on a number of legal implications but focus on the Fifth Amendment. Their conclusion is not without irony that current brain-reading technology may fall outside the scope of the privilege that the Fifth Amendment affords defendants precisely because of its current shortcomings, and only as long as these last: There can be no self-incrimination, the idea goes, so long as the technology has no access to the content of our brains.

The final theme appears to be a very workaday, legal matter. It is about whether and how neuroscience can be of assistance to our systems of criminal justice. But, in this domain, difficult scientific, psychological and philosophical questions persist alongside the legal ones. Let's assume that there is some controversy, in neuroscience, about whether a victim's brain injury is typically caused by an assault or not. How can judges and juries possibly understand the complex neuroscientific evidence? What is the current legal status of neuroscientific evidence: is it admissible or not according to the usual standards of admissibility? How useful is the evidence in fact? Joanna Glynn illuminates the legal context in England and Wales, in detail, analysing the extent to which neuroscientific evidence might be ruled inadmissible by virtue of concerns over reliability. The interesting possibility is explored that, in this regard, neuro-evidence might properly be treated more cautiously than other forms of evidence, where usually concerns about reliability go to weight

[1] John Searle, 'What your computer can't know', *The New York Review of Books* (2014). Along similar lines see: Searle, 'The Chinese room' in R. A. Wilson and F. Keil (eds.), *The MIT Encyclopedia of the Cognitive Sciences* (Cambridge, MA: MIT Press, 1999).

rather than to admissibility. Glynn explores some of the reasons for this caution, noting, for example, the suggestion that adversarial systems encourage medical experts to express their views in terms of certainties, even where the grounding science is much less certain.

Criminal courts need evidence; they need it, inter alia, to assess responsibility. What is the relationship between an offender's brain and that person's responsibility for a criminal action? Are some defendants less responsible, or more, than others who perform the same crime, and if so does it matter whether the relevant mitigating or aggravating factors are connected to brain states and structures or to societal or familial influences, for instance? The question matters a great deal for those whose behaviour departs from expected norms. Raynor is in no doubt that neuroscience leaves our common-sense and ordinary-language use of responsibility and culpability largely intact. Raynor asks for greater conceptual clarity about the meaning of the term *responsibility*, particularly in the context of criminology, penology and rehabilitation theory, noting that such clarity will enable a better understanding of how psychological reports might be relevant to sentencing; of how cognitive behavioural therapy affects offenders' choices; and about how social and individual factors can lead to abstention from offending. He makes the case that some circumstances that commonly occur in criminal justice call for a realistic view of responsibility based on how far people could reasonably have been expected to exercise control over their behaviour.

Neuroscience might tell us that some defendants are behaviourally 'normal' but, on a particular occasion, fail to act in accordance with the norm. Others are behaviourally 'abnormal', with a reduced capacity to exercise control and experience fear. One such 'abnormality' often manifest in the context of criminal offending is psychopathy. These kinds of questions emerge: In what way is a psychopath mentally different from a nonpsychopath? Are these differences attributable to structural brain 'abnormalities'? What does it mean to lack the capacity to conform one's behaviour to social, legal or moral norms, and what does it mean for the brain to evidence such a lack of capacity? Is it morally more blameworthy or less to offend in virtue of a lack of capacity than to offend while having a capacity and failing to exercise it? Marion Godman addresses these and related questions. She examines the growing challenge from scientists and philosophers to view deficits in empathy and moral cognition as reasons to excuse rather than to blame psychopaths. In responding to the challenge she notes that offending by psychopaths often takes the form of instrumental, rather than the more commonplace reactive, aggression.

Godman explains that the nature of their aggression indicates both that psychopaths exercise choices about their behaviour and that early intervention, to prevent offending, is justified. Her contribution is original in that Godman considers the features of psychopathy against the notion of fairness. Without committing to a particular view, Godman considers the interesting possibility that whether psychopathic tendencies are inculpatory or not, there might be a duty, falling on society, somehow to compensate the psychopath for a condition, and its associated failings, that, after all, he has not chosen to have.

In the final chapter, Elizabeth Shaw examines the potential use in criminal justice systems of direct biomedical interventions designed to alter the brain states or behaviour of psychopaths. Shaw examines two objections to the interventions: (a) the idea that for reasons of principle it would be futile to attempt to alter psychopathic traits biomedically and (b) the claim that there is something about the nature of psychopathy, which means that it could *never* be ethically permissible to offer such offenders treatments, that carry significant risks. Shaw rejects both objections and proposes a list of preconditions to be satisfied before neurointerventions could permissibly be given.

Many of the chapters in this book began life at a neurolaw conference hosted by Swansea University in December 2014. The debates were lively, and we learned much from each other's disciplinary approaches. This is a forum for integrating the various perspectives, for exploring what neuroscience can offer to law and for considering how the institution should respond to the undoubted enlightenment neuroscience provides, the challenges it presents, and the limitations it has yet to overcome.

PART I

Conceptual Disputes

Brains as the Locus of Responsibility?

1

Neuroscience and the Explanation of Human Action

DENNIS PATTERSON[1]

Introduction

Owing to the manifold advances in neuroscience, it is now widely suggested that neuroscientific explanations of human behaviour are actually superior to our ordinary ways of characterizing human action (typically referred to as 'folk psychology').[2] The implications of this for law are straightforward. In place of the usual forms of understanding and explaining human action (intentions, reasons and causes), we can now resort to the explanatory apparatus of cognitive neuroscience (electricity, synapses, neurons, the passing of blood through certain areas of the brain). Succinctly put, reasons for action will be replaced by causal explanation[3]: scientific explanation will displace our ordinary accounts of human action by reducing them to facts about brain states.

[1] My thanks to Bebhinn Donnelly-Lazarov for excellent comments and suggestions on a draft of this chapter. Thanks to my research assistant, Sofie Møller, for excellent editorial and substantive suggestions.

[2] Stephen J. Morse, 'Determinism and the death of folk psychology: Two challenges to responsibility from neuroscience' (2008) 9 *Minnesota Journal of Law, Science & Technology*, 1–36.

[3] When I use the term *cause* or *causal*, I mean 'efficient' causation. Aristotle identified four different types of causes: Efficient, Formal, Material and Final/Teleological. 'Efficient' causal explanations of why some event occurred typically describe some other preceding event that brought about the first event (e.g., 'the pins fell over because the bowling ball struck them' or 'the window broke because the baseball struck it'). By contrast, 'formal' explanations explain why something is so by reference to the nature or essence of the thing (e.g., 'It is a triangle because it has three sides' or 'He is a bachelor because he is an unmarried male'); 'material' explanations refer to the material constitution of something (e.g., 'It conducts electricity because it is made of copper'); and 'final/teleological' explanations refer to why something is so by reference to its purpose (e.g., 'He did it for her sake.'). For an illuminating overview, see P. M. S. Hacker, *Human Nature: The Categorical Framework* (New York: Oxford University Press, 2007), pp. 181–198.

In this chapter, I will argue that the aspiration to replace our current modes of explanation for action with a scientific account is untenable. Specifically, I will maintain that human action can never be properly understood nor explained solely in causal terms. In the course of my rebuttal to those who nurture the contrary view, I shall provide specific examples of the arguments for the proposition that scientific explanation of human phenomena is superior to and can replace the forms of understanding of human actions presupposed by our social institutions, especially law. The arguments are philosophical in nature: they depend not at all on empirical claims about facts nor do they draw their strength from any source other than their reasonableness.

In addition to showing why the causal account of human action comes up short, I want to advance a positive account, one grounded in the idea that reasons are central to understanding and explaining human action. In the course of arguing for the primacy of reasons in the explanation of action, I shall argue that reasons are best understood as normative notions and exactly that which is beyond the reach of scientific explanations about causal connections, as normative connections are of a different kind. The third and final aspect of the argument is the claim that the normative character of reasons arises from human practices.

The Reductionist Position

The rallying cry of the cognitive reductionist is 'You are your brain'. Here is a representative example of the position:

> We feel as if we are uncaused causers, and therefore granted a degree of independence from the deterministic flow of the universe, because we are unaware of the deterministic processes that operate in our own heads. Our actions appear to be caused by our mental states, but not by physical states of our brains, and so we imagine that we are metaphysically special, that we are non-physical causes of physical events.[4]

There are at least three things to notice in this account of the nature of human action. First, everything in the universe is material and, as such,

[4] Joshua Greene and Jonathan Cohen, 'For the law, neuroscience changes nothing and everything' (2004) 359 Philosophical *Transactions of the Royal Society B: Biological Sciences* 1451, 1781.

subject to the physical laws of the universe. Second, there are no non-physical causes of actions or events: everything that happens in the world is the result of causal forces. Finally, our belief that we decide what we will do and what actions we take is just an exercise in self-deception. We are not in charge of our actions. We do not decide what to do, for that is a matter of forces largely unknown to us over which we exercise no control.

Of course, this perspective on human action implicates a wide variety of issues, including free will and responsibility.[5] If such a view of human nature proved to be accurate, it would have important public policy implications, not least for criminal law. Here is an expression of those implications:

> [W]e maintain that advances in neuroscience are likely to change the way people think about human action and criminal responsibility by vividly illustrating lessons that some people appreciated long ago. Free will as we ordinarily understand it is an illusion generated by our cognitive architecture. Retributivist notions of criminal responsibility ultimately depend on this illusion, and, if we are lucky, they will give way to consequentialist ones, thus radically transforming our approach to criminal justice. At this time, the law deals firmly but mercifully with individuals whose behaviour is obviously the product of forces that are ultimately beyond their control. Some day, the law may treat all convicted criminals this way. That is, humanely.[6]

The basic contentions of this point of view are clear: indeed, they are direct corollaries of the more general view stated earlier. Human action is the product of forces beyond the control of the individual. Hence, the view that we choose our actions is just an exercise in self-deception.

The law presupposes a rather different view of human action. We punish people for their bad actions, and law is one means for this. Owing to the fact that we have free will, the law punishes us for intentionally committed bad acts. But if all our actions are the result of causal forces beyond our control, the law punishes persons for acts they could not have otherwise chosen to refrain from. Thus, the criminal law is both unjust and inhumane.

[5] A recent collection of papers on this topic is Walter Sinnott-Armstrong, *Moral Psychology: Free Will and Moral Responsibility* (Oxford: Oxford University Press, 2014).

[6] Green and Cohen, note 3, 1784.

Philosophical Presuppositions

Before turning to the specifics of the view sketched earlier, a word should be said about the more general philosophical position that informs it. The view is philosophical naturalism.[7] Like many philosophical notions, the precise meaning of the term is debated. Here I will focus on the version called 'physicalism' because this is the philosophical theory that underlies much cognitive neuroscience. Physicalism has three core ideas, the first two of which it shares with most other naturalist positions:

1. Everything that exists (i.e., all that is real) is found in nature;
2. There is nothing 'supernatural'; and
3. Natural science (i.e., scientific method) is the best way to investigate any phenomenon.[8]

A version, perhaps a 'stronger' version, of naturalism is physicalism. Galen Strawson explains:

> I'm a naturalist, an out-and-out naturalist, a philosophical or metaphysical naturalist, a naturalist about concrete reality. I don't think anything supernatural or otherwise non-natural exists.
>
> You can't classify anything as supernatural or non-natural until you have a substantive conception of the natural in relation to which something can be classified as non-natural. I do have one: I take it that concrete reality – anything that exists in space-time – is entirely *physical*. I'm a physicalist naturalist, and I think metaphysical naturalism is the same thing as physicalism as I've just defined it (I'm putting ethics aside). There are, however, important questions to be raised about what this amounts to. They're old questions, but they haven't received enough attention recently. One result of this is that many – probably most – philosophers who call themselves naturalists are in fact extreme anti-naturalists. They're false naturalists – *not*uralists.[9]

[7] I think it uncontroversial to say that 'naturalism' is a somewhat fraught term. There are varieties of naturalism, and of course, not every naturalist identifies with all variations of the perspective. I try to identify the various approaches to naturalism in the pages that follow.

[8] For an excellent introduction, see David Papineau, 'Naturalism' in Edward N. Zalta (ed.), *Stanford Encyclopedia of Philosophy*, https://plato.stanford.edu/, last accessed, 9 September 2014. See also Geert Keil, 'Naturalism' in Dermot Moran (ed.), *The Routledge Companion to Twentieth-Century Philosophy* (London: Routledge, 2008), pp. 254–307 and Michael S. Pardo and Dennis Patterson, *Minds, Brains, and Law* (Oxford: Oxford University Press, 2013), pp. 28–30.

[9] Galen Strawson, 'Real naturalism' (2013) 35 *London Review of Books*, 18, 28–30.

Physicalist naturalism clearly differs from garden-variety naturalism in one important respect: reductionism, the third characteristic mentioned earlier. The garden-variety naturalist does not dispute the importance of reality nor scientific method.[10] But unlike Strawson, garden-variety naturalists prescind from physicalist reductionism. Strawson, and many others, are critical of these forms of naturalism. The reason he is critical is their failure to embrace the reductive dimension of physicalism.

In the realm of neuroscience, the reduction sought is that of mind to brain, mental to physical. As Nobel laureate Francis Crick put it, '[t]he scientific belief is that our minds – the behaviour of our brains – can be explained by the interactions of nerve cells (and other cells) and the molecules associated with them'.[11]

When it comes to mind–brain reductionism, no one is more emphatic than Patricia Churchland. She writes,

> Brains incessantly make decisions: some pertain to the immediate future, and others to the distant future; some are trivial, and others are momentous. . . . The brain is a causal machine. Or, perhaps more accurately, given everything that is so far known in neuroscience, it is very probable that the brain is a causal machine. . . . Available evidence indicates that the brain is the thing that thinks, feels, chooses, remembers, and plans. That is, at this stage of science, it is exceedingly improbable that there exists a non-physical soul or mind that does the thinking, feeling, and perceiving, and that in some utterly occult manner connects with the physical brain. Broadly speaking, the evidence from evolutionary biology, molecular biology, physics, chemistry, and the various neurosciences strongly implies that there is *only* the physical brain and its body; there is no non-physical soul, spooky stuff, or ectoplasmic mind-whiffle.[12]

I think this is about as strong a statement for reductionism as one can find in the literature. Churchland wants to reconceive everything from ethics to epistemology from the point of view of the brain. All manner of 'mind talk' can be reduced to what the brain does. As she states explicitly: the brain thinks, makes decisions and comes to

[10] Richard Rorty, for example, described himself as a naturalist. See Richard Rorty, *Philosophy and the Mirror of Nature* (Princeton, NJ: Princeton University Press, 1979), p. 387.

[11] Francis Crick, *The Astonishing Hypothesis: The Scientific Search for the Soul* (London: Touchstone Books, 1995), p. 7.

[12] Patricia Smith Churchland, 'Moral decision-making and the brain' in Judy Illes (ed.), *Neuroethics: Defining the Issues in Theory, Practice and Policy* (Oxford: Oxford University Press, 2005), pp. 3–5.

normative conclusions. If we want to know what thinking is, study only the brain. But is this sort of reductionism plausible? We turn now to that question.

Reductionism

Recently, in the pages of *The New York Review of Books*, Colin McGinn debated the merits of mind–brain reductionism, first with French neuroscientist Jean-Pierre Changeux and shortly thereafter with Patricia Churchland. In his book,[13] Changeux argues that aesthetic experience can best be understood from the neuronal perspective. That is, if we truly want to understand the nature of aesthetic experience, we need to look at what is going on in the brain when we are looking at art.

Here is what McGinn says about Changeux's claim to be providing an account of aesthetic experience:

> Suppose Changeux agrees that he is not really discussing beauty itself, as an attribute of artworks, but rather our perception of beauty – the act, not the object. So he is discussing an expressly psychological subject – aesthetic experience. Now he tells us about the neuronal correlates of such experience, which there must surely be (this is the sliver of truth I referred to earlier). Does it follow that he is discussing aesthetic experience? Not unless aesthetic experience reduces to its neural correlates. For that to be so the mind must in general reduce without residue or remainder to the brain; then he can claim to replace mind talk with brain talk. Certainly, such a view has been maintained – it is the view usually labeled 'materialism' – but Changeux needs to persuade us that this philosophical position is correct, which he signally fails to do. He seems to think that if he can identify brain correlates for a mental state, he has thereby shown that there is nothing to the mind beyond the brain.[14]

McGinn puts his finger directly on the problem that eludes Changeux: no amount of empirical evidence is going to sustain what is, at bottom, a philosophical theory. Changeux's claim is that materialism is the best explanation of aesthetic experience. This is a metaphysical claim, not an empirical one. Changeux tries to traverse the explanatory gap between

[13] Jean-Pierre Changeux, *The Good, the True and the Beautiful: A Neuronal Approach* (New Haven, CT: Yale University Press, 2012).

[14] Colin McGinn, 'What can your neurons tell you?' (July 11, 2013) *The New York Review of Books*, (reviewing Jean-Pierre Changeux, *The Good, the True and the Beautiful: A Neuronal Approach* (2012)).

the fact of neuronal activity and the correct conceptual characterization of that activity. His argument is a non sequitur.

Science does not, indeed cannot, replace philosophy.[15] McGinn makes this point by mentioning that 'even a Cartesian dualist believes in brain correlates for all mental phenomena – he just denies that these are mental phenomena'.[16] The difference between the materialist and the Cartesian is solely a philosophical one. Changeux seems not to notice that he is propounding a metaphysical thesis.

What are the merits of materialism as a metaphysical position? Recall Galen Strawson's view articulated earlier: nothing exists unless it is material. Is this plausible? There is reason to doubt. Consider three things we find in the world: laws, baseball and poetry.[17] None of these is material. Of course, each has a material component. Laws are written in ink and found in books. But no one would say the law is the ink or the book.

Remember that our interest lies in explanation. The reductionist claims that we can do away with all mental talk and focus solely on what goes on in the brain when we want to understand human beings and their actions. We have no need of epiphenomenal mind talk. Again, this claim is extremely dubious. A number of philosophers have made the point that we cannot explain human action solely from a material point of view. Two arguments come to mind.

In 'Interpretation and the Sciences of Man',[18] Charles Taylor makes the case for a fundamentally hermeneutical explanation of human action. His point is relatively simple: context is ineliminable. Taylor's example was voting: how could you understand what someone was 'doing' or how could you explain the action when all you took account of was the person ticking a box? His point is that you could not explain what the person was doing without resort to their reason for ticking one box rather than another. To explain that, you would of necessity have to reach far beyond the content of the person's brain. Taylor's point is that by looking at the material context alone, you cannot explain human action.

[15] Churchland is positively exasperated at this aspect of McGinn's position. See Patricia Churchland, 'Of brains & minds: An exchange' (June 19, 2014) *The New York Review of Books*

[16] McGinn, note 13.

[17] The examples were found in M. R. Bennett and P. M. S. Hacker, *Philosophical Foundations of Neuroscience* (Malden, MA: Blackwell, 2003), pp. 358–359.

[18] Charles Taylor, 'Interpretation and the sciences of man' (1971) 25 *The Review of Metaphysics*, 1, 3–51.

Consider a different example: affixing a signature. This same motion of the hand can be at least five different actions: an autograph, executing a promissory note, executing a trust indenture, witnessing a will or signing a death certificate.[19] It is difficult to see how the same motion of the hand could be understood solely with knowledge of neural firings. At a minimum, these different actions require attention to both individual intentions as well as context. For reductionism to succeed as explanation, it would have to describe these disparate actions without resort to the intentional and contextual elements. This seems an impossible task.

Correlating Mental States with Brain States

When it comes to the relationship between mental states and brain states, some scholars have faith in the notion that we will have a better understanding of the human mind if we can identify the neural correlates of mental states.[20] Claims to the effect that this or that 'activity' takes place in a certain area of the brain are the basis for further (speculative) claims to the effect that the brain performs different functions in different areas of the brain.[21]

It is fallacious to think that someone has to be in a particular (neural) state in order to have a particular thought or mental process (e.g., moral reasoning). The reason is one of criteria. As Bennett and Hacker put it: 'the criteria of identity for mental states, events, and processes differ from the criteria of identity for neural states, events and processes'.[22]

Anthony Kenny supplies a good counterexample to the claim that correlation of brain states with mental states tells us something about the criteria of identity for mental states. In the course of an extended commentary on Chomsky's claim that the human mind exhibits a 'structure', Kenny draws this analogy:

> The conceptual points that I have been making can be made about pocket calculators no less than about human beings. My calculator works out the square root of 123456789. In a flash, there comes the answer 11111.11106. Between my pressing the square-root key and the display appearing,

[19] The example comes from Bennett and Hacker, note 16, p. 360.

[20] For a superb discussion of the conceptual issues facing the correlation exercise, see Gabriel Abend, 'What are neural correlates neural correlates of?' (2017) 12 *BioSocieties*, 3, 415–438.

[21] This approach is particularly well developed in the work of Joshua Greene. See Joshua Greene, *Moral Tribes: Emotion, Reason, and the Gap between Us and Them* (New York, NY: Penguin Press, 2013).

[22] Bennett and Hacker, note 16, pp. 360–361.

> complicated events took place in its electronic innards. Those events, whatever they were, could have taken place in a different calculator doing a different job; and a different calculator doing the same job might well have taken an electronically totally different route. Moreover, the hardware might have been taken out of the case, separated from the input keys and the output display. Whatever electronic events then took place inside it would not have been the working out of the square root of 123456789. For in the sense in which calculators can work out square roots for us, it is only whole calculators, and not portions of their electronic anatomy, that can do the working out.[23]

The moral of Kenny's example is twofold.[24] First, there is no one-to-one correspondence between hardware and software. How one computer makes the square-root calculation can be completely different from another. The same thing – computing the square root of 123456789 – can be done in myriad ways. Second, if the input keys had been separated from the hardware, it would not matter what went on inside the computer: nothing would constitute 'computing the square root of 123456789'. In short, there is no reason to believe that there exists any informative connection between how a task is accomplished and what doing the task in question consists in. Neural states – no matter how well identified – are not criterial. The most they can be is correlated with our activities.[25]

The Modern Theory of the Will

In one way or another, each of the topics just discussed relies on a picture of mental causation that John Hyman has called 'The Modern Theory of the Will'.[26] He describes it this way: 'The modern theory represented voluntary action, intentional action, and action as such – i.e. action that can be imputed to an individual as the agent – by means of a single picture: desire (or the "uneasiness" of desire) causes the mental action of volition, which in turn causes motion in our limbs'.[27]

Wittgenstein and Ryle each make an appearance in Hyman's narrative. However, Hyman maintains, neither went far enough in their critique of this dominant picture of human action. Hyman sees human action as

[23] Anthony P. Kenny, 'Language and the mind', repr. in *The Legacy of Wittgenstein* (Oxford: Blackwell, 1984), p. 142.

[24] See Bennett and Hacker, note 16, p. 361, n. 11.

[25] For discussion, see Pardo and Patterson, note 7, pp. 8–12.

[26] John Hyman, *Action, Knowledge, and Will* (Oxford: Oxford University Press, 2015).

[27] Ibid., p. 12.

composed of four elements: physical, psychological, intellectual and ethical.[28] There are reasons to be optimistic that Hyman's taxonomy will enable us to make real progress in understanding human action in ways not previously possible with the modern theory's emphasis on an empirical or causal model of action.

Let me give one example of how Hyman's four-part typology enables us to see something in human action that has been obscured by the Modern Theory of the Will. Consider the concept of 'voluntary act'. This notion is central to law and legal theory. John Austin described it as follows: 'A voluntary movement of my body, or a movement which follows a volition, is an act'.[29] This definition has been signed onto by leading judges and theorists both in the UK and the United States.[30] It is confused.

Hyman does an excellent job in showing the reach of the confusion among philosophers in understanding the relationship between action and voluntariness.[31] His key insight is that we assess voluntariness as action that is not accompanied by ignorance or compulsion (i.e., 'negatively'). Voluntariness is a quality of an act that can be defeased if the act is done out of ignorance or compulsion. Hyman's best example of this is compulsion or duress.[32]

[28] Ibid., p. 4.

[29] John Austin, *Lectures on Jurisprudence, or, the Philosophy of Positive Law* (London: J. Murray, 1869), lecture XVIII.

[30] Hyman writes:

> The American jurist and Supreme Court justice Oliver Wendell Holmes writes: 'An act is always a voluntary muscular contraction and nothing else . . . An act implies a choice'. The English jurist D. A. Stroud defines a 'simple' act as 'a voluntary movement of the body', and a complex act as 'a series . . . of simple acts and their consequences'. Glanville Williams defines an act as 'a willed bodily movement', adding, 'Since every act is by definition willed, there is no need to call it voluntary'. R. W. M. Dias defines an act as a 'voluntary bodily movement', adding that voluntariness is 'a criterion of responsibility and connotes controllability of the action in question'. Michael S. Moore defines an act in the same way as Glanville Williams, as a 'willed bodily movement', and states (now following Mill) that 'human acts are themselves a compound of other events, consisting of a volition causing a bodily movement'.
>
> Hyman, note 25, p. 11.

[31] Ibid., p. 77.

[32] What follows is a compressed version of my discussion of this topic in Dennis Patterson, 'Rethinking duress' (2016) 7 *Jurisprudence*, 3.

Since Elizabeth Anscombe wrote on the subject, the conventional wisdom has been that an act done intentionally is done voluntarily.[33] But an act done under duress is the product of compulsion. The act can be done 'intentionally' but if it is the product of a threat, it cannot be voluntary. The reason it cannot be voluntary is that actions done under duress typically are motivated by threats that make acquiescence unpalatable. A simple example is the threatening gunman. When you hand over your wallet in response to the threat of violence, you intend to give the thief your wallet. But it hardly seems correct to say you did so voluntarily.

Much more can be said both about voluntariness and the effect of compulsion. The small point I want to make here is that Hyman's critique of the Modern Theory of the Will opens up the real possibility of rethinking the relationship between the intentional and the ethical. It does seem to be an error to equate the intentional with the voluntary. In that, Hyman is right.

From Causes to Reasons

To this point, I have been concerned with causal approaches to the explanation of human action. I have endeavoured to show why at least some forms of causal explanation of human action are ill conceived or worse. I shall now shift my focus from critique to explication. Specifically, I will make the case that the best way to understand and explain human action is from the point of view of reasons, not causes. The arguments are motivated by the belief that an understanding of behaviour is only possible at the level of reasons for action. It is only at this level that the normative dimension of behaviour is properly understood.

Reasons versus Causes

Mental life – e.g., believing, intending, deciding – is lived in the space of reasons and not in the space of causation.[34] To see the difference, consider the following examples:

1. A doctor taps on my knee, and my leg moves.
2. I watch a sad movie and begin to cry.[35]

33 See G. E. M. Anscombe, *Intention* (Oxford: Blackwell, 1957).

34 Of course, Donald Davidson famously argued that reasons could be causes. See Donand Davidson, 'Actions, reasons, and causes' (1963) 60 *Journal of Philosophy*, 685.

35 The examples are taken from Svend Brinkmann, 'Mental life in the space of reasons' (2006) 36 *Journal for the Theory of Social Behaviour*, 1, 2–3.

Example 1 is a strictly causal event. The reaction of my leg to the doctor's tap can be fully explained in causal (i.e., material) terms: a case of reflexive movement. I had no reason for moving my leg: reason (i.e., my reason) was never an aspect of my reaction to the force of the doctor's hammer on my knee.

In the second case, matters are more complicated. First, a straightforward question: is the movie the cause of my sadness? It appears that it is. But this appearance is deceiving. Unlike Example 1, Example 2 involves a mental event. In Example 2, I can offer a reason why I am sad. Additionally, I can be asked to justify my reaction to the film. Perhaps I am mistaken in believing there is a reason to cry. I may have misunderstood the filmmaker's intentions, and when these are brought to my attention, I change my view. It is in the role of justification of my reaction that we find the seeds of the key difference between the first (causal) case and the second (reason) case. The difference is a matter of normativity. In the first case, it is impossible to say 'It was wrong of me to move my leg', whereas in the second, it is perfectly possible that my reaction was inappropriate or wrong in some way. The normativity of the mental is the basis for my claim that causal explanation can in no way displace the realm of reasons.[36]

The Mental as Moral

Wilfrid Sellars's classic *Empiricism and the Philosophy of Mind* (1956) showed how some of our most basic concepts are, at bottom, normative. He said this about knowledge:

> The essential point is that in characterizing an episode or a state as that of knowing, we are not giving an empirical description of that episode or state; we are placing it in the logical space of reasons, of justifying and being able to justify what one says.[37]

Knowing is a mental (and not a physiological) phenomenon. Sellars's point was that as a mental phenomenon, claims to knowledge require justification. As Gettier showed, it is not enough that a belief happens to

[36] The foregoing relies on the explication in ibid., 2–3.

[37] Wilfrid Sellars, *Empiricism and the Philosophy of Mind* (Cambridge, MA: Harvard University Press, 1997), p. 76.

be in accord with the facts: it may well be that a particular claim is nothing more than a lucky guess.[38]

Much the same can be said about cognition, perception, reasoning, problem solving and memory. These are normative because we can demand reasons of anyone making a claim in any of these realms. They are, as it were, 'achievement terms', which, as such, presuppose a normative difference between 'correct' and 'incorrect'. We can say the same about intentional action: it too is normative because it calls for 'explanation of behaviour in terms of agential reasons'.[39]

The key to unlocking the difference between causes and reasons is found in the normative nature of reasons. Mental phenomena are normative because agents act within a space of reasons. MacIntyre is worth a lengthy quote on this point:

> In every culture emotions and desires are norm-governed. Learning what the norms are, learning how to respond to the emotions and desires of others, and learning what to expect from others if we exhibit certain types and degrees of emotion or desire are three parts of one and the same task. What the norms govern is a set of three-term relations: *a particular type or degree of emotional reaction or of aspiration to fulfill a desire* is justified in *such and such a range of situations* when exhibited by *such and such a type of person filling such and such a particular type of social role.* If I do not know how the relations between these are norm-governed, then I will not understand, for example, why others react as they do to what they judge to be my unjustified anger or what it is that makes their indignation at my anger justified and so unacceptable to people in general in a way that my anger is not. . . . Thus, to exhibit a particular pattern of emotions and desires, to treat them as appropriate or inappropriate in one situation rather than another, is always to reveal a commitment to one set of justifying norms rather than another. For the justifying norms which govern both emotions and desires and their interpretation either embody or presuppose a rank ordering of goods and evils. Hence at any particular stage in the historical development of any particular culture the established patterns of emotions, desire, satisfaction, and preferences will only be adequately understood if they are understood as giving expression to some distinctive moral and evaluative position. Psychologies thus understood express and presuppose moralities.[40]

[38] Edmund L. Gettier, 'Is justified true belief knowledge?' (1963) 23 *Analysis*, 6, 121–123.

[39] Hacker, note 2, p. 217.

[40] Alasdair MacIntyre, *Whose Justice, Which Rationality?* (Notre Dame, IN: University of Notre Dame Press, 1989), pp. 76–77.

The Explanatory Power of Reasons

I should like to draw together three themes I identified earlier. First is the hermeneutic dimension of the explanation of human action. Second is the explanation of action by appeal to reasons. Finally, I shall return to the idea of the normative dimension of reasons.

The form of explanation I want to argue for in place of the efficient-causation conception is variously identified as teleological,[41] or final/purposive.[42] With a nod to Michael Thompson,[43] Eric Marcus describes and identifies the core feature of explanation by resort to reasons as 'explanation of part of an event by appeal to the whole of which it is a part, where the former is for the sake of the latter'.[44]

In explaining an action, we invariably look for the reason an agent acted as she did. We may inquire of the agent the reason for her action, but this is not a necessity; indeed, it is often the case that a reason for action is inferable from the context (e.g., someone waving (for a taxi) on a New York City street). By contrast, we might want to explain why someone cried 'Out' in a tennis match. The utterance is best explained not by the brain actuating the vocal cords but by appeal to the rules of tennis. Many explanations have this normative character, as they involve appeal to rules or standards in explaining why someone acted as they did.

As Michael Pardo and I have argued,[45] in explaining an action in terms of the reasons for it, the action may be explained in terms of the purpose for which it was carried out. A baseball player moves as he does (e.g., bunting) '*in order to* move to [sic] runner on first base to second base'.[46] The player's explanation may be true or false: if the action is performed for the reason given, then the explanation is true.

[41] Eric Marcus, *Rational Causation* (Cambridge, MA: Harvard University Press, 2012), p. 93.

[42] Hacker, note 2, p. 179.

[43] Michael Thompson, *Life and Action* (Cambridge, MA: Harvard University Press, 2008), part 2.

[44] Eric Marcus, note 40, p. 92.

[45] Michael S. Pardo and Dennis Patterson, 'Morse, mind, and mental causation' (July 22, 2014) 11 *Criminal Law and Philosophy*, 1, 111–126. We went on (in a footnote) to add: 'compare such an explanation with the mens rea requirement of acting 'purposely' under the Model Penal Code. The explanation of the baseball player's conduct meets the definition of acting purposely in bunting – i.e., it was his 'conscious object to engage in conduct of that nature' or 'to cause such a result' – even though the explanation does not depend on efficient mental state causation.

[46] Pardo and Patterson, note 7, at 11.

Against this view, one might say that what I have provided is just a causal explanation in different words. Even if one wanted to say 'the player's bunt caused the runner to advance', that use of 'cause' is not efficient (like a bowling ball hitting pins) but more in the nature of an 'event-as-a-whole' explanation of the parts.[47]

A Normative Conception of Practices

The normativity that suffuses the mental and supplies the grounds for reasons in the justification of action is not a product of the individual.[48] Mental life is not lived at the level of the individual; rather, it is suffused in the practices of everyday life. Of course, people think and have their own thoughts. But it is only possible for people to think 'because their activity is framed in a space of reasons in which they are susceptible to normative standards that do not arise from individual mentality'.[49]

Normativity has a social basis. There is no better argument for this than Wittgenstein's rule-following arguments in the *Philosophical Investigations*.[50] The literature of 'practice theory' is deep with arguments grounded in Wittgenstein's rule-following remarks.[51] Brandom, for example, builds his theory of linguistic competence as deontic scorekeeping out of a not uncontroversial reading of the role of interpretation in

[47] Peter Hacker explains the contrast between causes and reasons thus:

> A mere muscular contraction is typically insufficient to determine what act was performed, and its neurophysiological explanation cannot explain whatever act was performed in or by moving voluntarily or intentionally.
>
> Reasons, by contrast, explain acts and omissions. It is true that the agent's wanting to W or to attain G is commonly a factor in explaining his V-ing. But this explains the agent's action, not its constitutive movements. If we are puzzled by a chess-player's gambit, we want an explanation of his move, not of his movement. Neural events in the brain may explain how it is that the chess-player is able to move his hand. If they did not occur as normal, he would not be able to move or control the movements of his hand. But neural events cannot explain his move – only those principles of chess strategy that he was aware of can do that. Nor can they warrant the redescription of his action in terms of its normative consequences – for example, that he checked his opponent.

Hacker, *Human Nature*, note 2, p. 230.

[48] And not, in particular, the individual 'mind-brain'. [49] Brinkmann, note 34, 9.

[50] Ludwig Wittgenstein, *Philosophical Investigations* (Chichester: Wiley-Blackwell, 2009), paras. 201–202.

[51] See, for example, Theodore R. Schatzki, *Social Practices: A Wittgensteinian Approach to Human Activity and the Social* (Cambridge: Cambridge University Press, 1996); Charles Taylor, *Philosophical Papers: Volume 2. Philosophy and the Human Sciences* (Cambridge: Cambridge University Press, 1985).

Wittgenstein's rule-following argument.[52] Brandom's discussion is quite useful for present purposes. When it comes to how best to understand practices, Brandom draws a distinction between 'regulist' and 'regularist' approaches[53] to understanding practices.[54] The distinction is advanced to better understand the normativity of practices, specifically discursive practices.

In our discursive practices, we have resort to the words *correct* and *incorrect, right* or *wrong*. How are we to understand these terms of appraisal? The regulist maintains that it is in virtue of some rule that a performance is characterized as correct or not. The regularist, by contrast, locates the ground of appraisal in what the community does, its behaviour. As Brandom puts it:

> The simple regularity approach is committed to identifying the distinction between *correct* and *incorrect* performance with that between *regular* and *irregular* performance. A norm implicit in a practice is just a pattern exhibited by behaviour. To violate that norm to make a mistake or act incorrectly according to that norm, is to break the pattern, to act irregularly.[55]

Rules do not apply themselves (regulism). There must be more to correct and incorrect performance than communal agreement.[56] But what more? For Brandom, the additional element is 'mutual accountability'.[57] In discursive practices, participants make claims and respond to the challenges of other participants, all in an ongoing context of mutual accountability. As Brandom put the matter: 'we can envisage a situation in which *every* social practice of [a] community has as its generating response a performance which must be in accord with another social practice'.[58]

Joseph Rouse provides a useful elaboration of Brandom's argument. He identifies three salient aspects of Brandom's account of normativity:

[52] See Robert Brandom, *Making It Explicit* (Cambridge, MA: Harvard University Press, 1994), 199–201.

[53] A succinct explication of these terms is found here: http://duncanlaw.wordpress.com/2011/01/05/regulism-and-regularism/.

[54] Brandom, note 51, pp. 20–21.

[55] Ibid., pp. 28–29.

[56] The best account of this aspect of the argument is found in Meredith Williams, *Wittgenstein, Mind and Meaning: Towards a Social Conception of Mind* (London: Routledge, 1999).

[57] See Brandom, note 51, p. 713.

[58] Ibid., pp. 189–90.

1. The bounds of a practice are identified by the ways in which its constitutive performances bear on one another, rather than by any regularities of behaviour or meaning that they encompass;
2. These patterns of interaction constitute something at issue and at stake in their outcome; and
3. Normativity involves a complex pattern of interrelations among performances through time. Such performances are normative when they are directed towards one another as mutually accountable to common stakes.[59]

Reasons for Action: Who Decides?

As previously mentioned, when we ask a person 'Why did you V?' a sincere answer serves as a justification. But is an agent an authority for her actions? Is there a first-person authority? No. Even when an agent sincerely offers a reason for action, that justification might be mistaken (think of the sad movie example given earlier). The agent's reasons have to make sense as an explanation of her action. If they do not, then the explanation will be rejected. Even if the agent sincerely believes what she is saying, the explanation can be false.[60]

In the language of Brandom, the agent offers an explanation of her conduct for which she takes responsibility. Taking responsibility means participating in a practice of reason-giving that features accountability and assessment. It may be that there is more than one narrative to explain our behaviour. Here it is difficult to say that one account is correct or not without due appreciation of the normative character of those terms.

Conclusion

Can reasons provide a better account of human action than causes? This is the central question of this chapter. There are two lines of assessment. First are the weaknesses in the causal account of action.[61] Second, does the reasons approach provide a sufficiently granular account of action such that it can be a viable alternative to the causal account? I suggest it does.

[59] Joseph Rouse, 'Social practices and normativity' (2007) 37 *Philosophy of the Social Sciences*, 1, 49–53.

[60] For discussion, see Hacker, note 2, at 223–25.

[61] For this, see Pardo and Patterson, 'Morse, mind, and mental causation', note 44.

2

"Nothing but a Pack of Neurons": The Moral Responsibility of the Human Machine

MICHAEL S. MOORE*

I Introduction

The world-famous geneticist Francis Crick wrote in 1994 that the science of human behavior had arrived at what Crick called "the astonishing hypothesis":

> The Astonishing Hypothesis is that "You," your joys and your sorrows, your memories and your ambitions, your sense of personal identity and free will, are in fact no more than the behavior of a vast assembly of nerve cells and their associated molecules ..." You're nothing but a pack of neurons."[1]

In truth the general form of Crick's "astonishing hypothesis" has been advanced by virtually every kind of science of human behavior, be it academic psychology/psychiatry/genetics/ethology/artificial intelligence,

* This chapter was originally given to the School Seminar Series of the John Curtin School of Medical Research, Australian National University, Canberra, Australia, September 2014; then as one of the Public Lecture Series of the Center for Neuroscience and Society of the University of Pennsylvania, Philadelphia, Pennsylvania, November 2014; then to the Conference on Law and Neuroscience, Swansea University College of Law, Swansea, Wales, UK, December, 2014; then to the Conference on Philosophy, Law, and Neuroscience, European University Institute, Florence, Italy, June 2015; and finally to the Conference on Neuroscience and Responsibility, University of Gothenburg, Gothenburg, Sweden, November 2015. My thanks go to the participants at these presentations for the many comments I received that helped shape the chapter into its present form. Thanks also go to the 11 other participants from around the world (Peter Alces, Selim Berker, Richard Fumerton, Frank Jackson, Bill Lycan, Stephen Morse, Dennis Patterson, David Papineau, Adina Roskies, Katrina Sifferd, and Bob Van Gulick) who over a two day period in September, 2015, discussed the topic of this chapter at the Roundtable on Neuroscience, Metaphysics of Mind, and Responsibility, held at Galena, Illinois, and cosponsored by the University of Illinois Program in Law and Philosophy and the University of Iowa Philosophy Department; that set of discussions was also quite helpful in giving this chapter its current shape.

[1] Francis Crick, *The Astonishing Hypothesis: The Scientific Search for the Soul* (New York: Simon and Shuster, 1994), p. 3.

or something else. In general form the hypothesis is that persons are just a kind of machine, a kind of mechanical clockwork that is subject to forces determining its behavior and over which it has no control. The general skeptical implication of this view is the same, irrespective of whether the machine simile is cashed out in terms of instinctual energy monotonously seeking discharge in sexual or aggressive behaviors (Freud's "dynamic" and "economic" meta-psychologies),[2] environmental stimuli reinforcing behavioral responses without the mediation of a free, dignified, or autonomous self (Skinner's behaviorism),[3] the invasion into the brain of lesion-causing entities (such as the spirochete in general paresis, the paradigm of mental disease of nineteenth-century psychiatry),[4] there being a "vital balance" of psychic forces determining both healthy and unhealthy behaviors (Menninger's paradigm for mid-twentieth-century psychiatry),[5] our survival-oriented genetic makeup (genetics), two-valued switching mechanisms (digital computers), or something else. In all cases, the conclusion is the same: we are not the responsible agents we imagine ourselves to be in our prescientific conceit.

Despite its similarity to its forbearers in science, neuroscience has captured the popular imagination in a way its predecessors never did. This is not because the kind of challenge is different, because it is not. Rather, it is because of three factors. One, the science is better than some of neuroscience's predecessors. Neuroscience is "hard science" in a way that Freud's metapsychology, Skinner's behaviorism, and Menninger's psychic determinism could never hope to be. Two, neuroscience on its face has the potential to give a *complete* explication of the working of the human machine in a way that sciences focusing only on environmental or genetic factors alone could never realistically aspire to. As Josh Green and Jonathan Cohen put this point, neuroscience studies the "bottleneck" through which all environmental and genetic causes must operate.[6] Third, neuroscience deals with *immediate* causes of human behaviors,

[2] I discuss Freud's five meta-psychological viewpoints in Moore, "Mind, Brain, and the Unconscious," in P. Clarke and C. Wright, eds., *Mind, Psychoanalysis, and Science* (Oxford: Basil Blackwell, 1988).

[3] B. F. Skinner, *Beyond Freedom and Dignity* (New York: Knopf, 1973).

[4] See the discussion in Michael Moore, *Law and Psychiatry: Rethinking the Relationship* (Cambridge: Cambridge University Press, 1984), pp. 224–228.

[5] Karl Menninger et al., *The Vital Balance: Life Process in Mental Health and Illnesses* (New York: Viking Boks, 1963).

[6] Joshua Greene and Jonathan Cohen, "For the Law Neuroscience Changes Nothing and Everything," *Philosophical Transactions of the Royal Society of London*, Vol. 359 (2004), pp. 1775–1785.

not remote causes. Such proximate causation not only yields stronger intuitions of mechanism in the popular mind,[7] but it also affords a much more plausible reduction base for mental states of desire, belief, intention, and will, than (e.g.) "behavioral constructs" such as dispositions, tendencies, and the like, ever could. So although the basic challenge to our self-conception issued by neuroscience is the same as that issued by its predecessors, the popularly perceived force of the challenge is considerably heightened.

My first task in this introduction is to make more precise the nature of the challenge on which I here wish to focus. For in truth neuroscience (again, like its predecessors) mushes together several quite distinct forms of challenge to responsible agency. Consider this statement of the challenge by David Brooks, a journalist–consumer of neuroscience, summarizing the views of others (which he himself does not share):

> Human beings are nothing but neurons . . . Once we understand the brain well enough, we will be able to understand behavior. We will see the chains of physical causation that determine actions . . . We will see that people don't really possess free will; their actions are caused by material processes emerging directly out of nature. Neuroscience will replace psychology and other fields as the way to understand action.[8]

In other work,[9] I have sought to show how such statements mix together four (plus) distinct challenges to responsible agency. The first of these is the centuries-old challenge of determinism: if all choices and actions are caused by factors themselves not under the control of the agent, how could she have done other than she did, how can she be responsible?[10] The second of these is the equally ancient challenge of epiphenomenalism: irrespective of whether human choices are caused (determinism) or

[7] Spatiotemporal contiguity of cause and effect is one of Ernest Nagel's criteria for when an explanation is "mechanistic." Nagel, *The Structure of Science: Problems in the Logic of Scientific Explanation* (New York: Harcourt, Brace, and World, 1961).

[8] David Brooks, Op-Ed, *New York Times*, June 17, 2013.

[9] Michael Moore, "Responsible Choices, Desert-Based Legal Institutions, and the Challenge of Contemporary Neuroscience," *Social Philosophy and Policy*, Vol. 29 (2012), pp. 233–279.

[10] The deterministic challenge is examined in detail in Michael Moore, "Compatibilism(s) for Neuroscientists," in Enrique Villanueva, ed., *Law and the Philosophy of Action* (The Netherlands: Rodopi Philosophical Studies, 2014), pp. 1–59, revised and reprinted in Moore, "Stephen Morse and the Fundamental Psycho-Legal Error," *Criminal Law and Philosophy*, Vol. 10 (2016), pp. 45–89; see also Moore, "The Neuroscience of Volitional Excuse," in D. Patterson and M. Pardo, eds., *Philosophical Foundations of Law and Neuroscience* (Oxford: Oxford University Press, 2016), pp. 179–230.

not ("free will"), those choices themselves do not cause those acts that are the immediate objects of such choices; rather, both choices and the acts those choices putatively cause are merely epiphenomenal of one another, i.e., coeffects of a common cause in the brain.[11] The third challenge concealed within blunderbuss challenges such as that characterized by Brooks is what I call the fallibilist challenge: we do not have the privileged access to knowing what we are thinking or doing (or why) that we assume we have, and without such knowledge there can be no control, no agency, and no responsibility.[12]

In this chapter, I intend to address none of these three forms of the challenge to responsibility raised by neuroscience. Rather, I intend to focus on a fourth and distinct form of challenge, what I shall call the reductionist challenge: we are – at bottom, ultimately – nothing but pieces of dumb, mechanical clockwork, nothing but "a pack of neurons" transferring energy in mechanical ways.

Just what scientific insight this challenge is based on, and just what its morally skeptical inference from that insight actually is, are no easy matters to state. Attempting to do so will occupy us in the body of this chapter. What can be done here preliminarily is by way of negative characterization: the distinctive challenge to responsible agency based on reductionism is whatever challenge(s) is (are) left once one subtracts the challenges of determinism, epiphenomenalism, and fallibilism, from the overall challenge to responsible agency based on our machine-like nature.

Some think that nothing is left over, that reductionist physicalism has no sting for responsible agency once one removes the stings of determinism, epiphenomenalism, and fallibilism. This is the idea that it is only because physical systems are (1) deterministic systems, (2) systems without causally efficacious wills, and (3) systems lacking any privileged access to their own internal states, that such systems lack responsible agency. Remove these stings, the thought is, and the bare fact that we are at bottom machines has no negative implications for responsible agency.

That is not the hypothesis that guides this chapter. Rather, my hypothesis is that there is an independent skepticism made tempting by

[11] The epiphenomenalist challenge is examined in Moore, "Libet's Challenge(s) to Responsible Agency," in Walter Sinnott-Armstrong and Lynn Nadel, eds., *Conscious Will and Responsibility: A Tribute to Benjamin Libet* (Oxford: Oxford University Press, 2011), pp. 207–234.

[12] Freud's version of the fallibilist challenge is examined in Moore, *Law and Psychiatry*, chaps. 3, 7–10.

reducing all that we value about ourselves to mechanical processes.[13] The skepticism is independent in the sense that such skepticism would exist even if we had free will, causally efficacious wills, and perfect self-knowledge because of infallible privileged access to our own mental states. No doubt, as the old saying goes, the proof is in the pudding. Thus, this chapter.

I intend to proceed by pursuing three enquiries. First, what do we mean by reductionism, both between scientific theories and between kinds of entities? How is reductionism related to physicalism; how it is related to mechanism? Second, how strong is the neuroscientific case that mental states and actions (the two touchstones of moral responsibility) can be reduced to brain states and to brain-state-caused bodily movements; that psychology can be reduced to neuroscience? Is all the world – and/or all thought about the world – in some sense physicalistic? In some sense mechanistic? Third, if so, what are the implications (for moral responsibility and blameworthiness) that stem – uniquely stem by not relying on determinism, etc. – from the truths of reductionism/physicalism/mechanism? Each of the three succeeding parts of the body of this chapter corresponds to one of these enquiries.

II From What Mind/Brain Metaphysical Views Does the Distinct Reductionist Challenge Arise?

It is far from clear what the unique challenge is that arises from the view that we are machines ("nothing but a pack of neurons"). Partly this is due to a lack of clarity about just exactly what are the neuroscientific views about mind and its basis in the brain from which this challenge supposedly arises. So let me start there.

There are five forms of physicalism, each having its adherents in neuroscience. Each also seems to be the basis from which some form of morally skeptical challenge arises. Before charting their differences, first attend to what makes all five of them versions of physicalism. This is the common denial of metaphysical dualism about the mind.[14] A dualist, as

[13] My aim in this chapter is to thread the needle between thinking that no such independent skepticism is plausible – which I deny – and thinking that such independent skepticism is true – which I also deny. This is thus an exercise in constructing a position that appears to tempt others but with which I ultimately disagree.

[14] Idealism is also an enemy common to all five forms of physicalism, but so few scientists or philosophers today are idealists that I put aside and ignore this old philosophical chestnut.

I shall define him, believes that persons have two features that do not exist in the physical world, agency and a mental life. Human actions, it is said, have as their essence an agency created by a human willing. Such willing is primitively agential in the sense that no account can be given of it in nonagential terms. Only near synonyms – like "agent-causation"[15] – can preserve the uniquely agential status of actions. Mental states, it is further claimed, have two features not to be found in the physical world. The first of these is conscious experience, or "qualia." The thought is that nothing in the physical world can have the subjective experience of a conscious self.[16] Also (and second), some of the mental states so experienced have the property of being directed on an object: an intent is an intent *to do something*, a belief is a belief that *something is the case*, a desire is a desire that *something be the case*, etc. Such objects are not to be confused with physical objects, for nonexistent sailboats can be wished for as earnestly as can real sailboats. The trick is both to describe the nature of the objects of such mental states better than ascribing to them "Intentional Inexistence"[17] and to show how physical stuff like brain states can be *about* anything, including such objects.[18]

All five forms of physicalism deny that either actions (agency) or mental states (qualia, intentionality) exist in any special, sui generis, nonphysical world. Physicalism is thus at bottom the monistic, one-worlder view denying that there are separate modes, kinds, or categories

15 Richard Taylor, *Action and Purpose* (Englewood Cliffs, NJ: Prentice-Hall, 1965).

16 The bases for the modern discussion of the qualia-based argument for dualism is threefold: Tom Nagel's claim that it has to be like something to be an object before that object can be thought to have subjective experience (Nagel, "What Is It Like to Be a Bat?" *Philosophical Review*, Vol. 83 (1974), pp. 435–450); Frank Jackson's argument that those previously deprived of some experience such as seeing or hearing but who, zombie-like, can discriminate in their behavioral responses as well as sighted or hearing subjects, will learn something new when they see or hear for the first time (Jackson, "Epiphenomenal Qualia," *Philosophical Quarterly*, Vol. 32 (1982), pp. 127–136); and David Chalmers's argument that there is a hard problem of consciousness that remains unsolved even after neuroscientists solve all the "easy" problems of finding unique correlates of conscious experiences of various kinds in the brain (Chalmers, "Facing Up to the Problem of Consciousness," *Journal of Consciousness Studies*, Vol. 2 (1995), pp. 200–219).

17 Franz Brentano's revival of the well-known phrase of the Scholastics of the Middle Ages for expressing the peculiarities they found in the idea of an Intentional object. Brentano, *Psychologie vom Empirischen Standpunkt* (Leipzig, 1874), a selection translated in Roderick Chisholm, ed., *Realism and the Background of Phenomenology* (Glencoe, IL: Free Press, 1960).

18 "Aboutness" and meaning lead John Searle to his mysterious "primitive intentionality." Searle, "Minds, Brains, and Programs," *Behavioral and Brain Sciences*, Vol. 3 (1980), pp. 417–424.

of existence. The five forms of physicalism differ in how they relate actions and mental states to items (like bodily movements and brain states) that enjoy unproblematic existence in the physical world.

Reductionist Physicalism

To say what a physicalistic reductionist (hereinafter, referred to simply as a "reductionist") believes requires one first to be clear as to what a reductionist generically believes. Yet saying what reductionism is, in general, is itself no easy task, in part because the question by now is so encrusted with a half century of detailed philosophizing about it.

A common entry point into this philosophical thicket is with Ernest Nagel's notion of scientific reduction in his classic book, *The Structure of Science.*[19] Like Putnam and Oppenheim before him,[20] Nagel conceived of the sciences as existing in a kind of natural hierarchy, a hierarchy in which some sciences were more basic than others. On this view, a less basic science (such as psychology) could be reduced to a more basic science (such as neuroscience), but not vice versa. When such a reduction was successful, two sets of relations would hold. First, the statements making up the experimental laws and theories of the reduced science would be deductible from the statements making up the laws and theories of the more basic science. And second, the things referred to in the reduced science would be identical to things referred to in the more basic science.

Nagel's now famous example was the reduction of classical thermodynamics to statistical mechanics. The laws of the former about the interdependence of pressure, volume, and temperature of a gas (Charles' Law, Boyle's Law, Snell's Law on the diffusion rate of a gas through a porous membrane, etc.) could be deduced from the laws governing the movement of gas molecules in random motion. Such reduction of *laws* was possible only because there existed Nagel's "correspondence rules" (what Hempel called "bridge principles") that connected the terms of classical thermodynamics (e.g., "temperature") to the terms of statistical mechanics (e.g., "mean kinetic energy of molecules in random motion"); the most intuitive kind of connection of terms here was made via an

[19] Nagel, *The Structure of Science.*

[20] Paul Oppenheim and Hilary Putnam, "The Unity of Science as a Working Hypothesis," in Herbert Feigl et al., eds., *Minnesota Studies in the Philosophy of Science*, Vol. 2 (Minneapolis: University of Minnesota Press, 1958), pp. 3–36.

identity claim between the things to which the terms referred (e.g., "temperature of a gas *is* the mean kinetic energy of its molecules in random motion").

So construed, Nagel's kind of reductionism was committed to two kinds of relationships: that of material implications (or perhaps equivalence) between the statements made by two theories, and that of identity between the things referred to by the terms used in such statements. With some force-fitting of the labels, these have come to be known, respectively, as "epistemological" and "ontological" reductions.[21]

Notice that Francis Crick's and Daniel Brooks's statements of skepticism combine both of these kinds of reductive claims: ontologically, our selves – including all that each of us values about ourselves, such as our agency, our joys, sorrows, memories, personal identity, and sense of self – are "nothing but a pack of neurons," in the sense that they each are identical to certain patterns of neuronal activity. Epistemologically, "neuroscience will replace psychology" because the statements of the latter can be reduced to statements of the former.

Let us henceforth designate the view that actions and mental states can be identified with certain physical events and states, "classically reductionist physicalism." We next need to explore less classical kinds of physicalism, each of which is in its own distinct sense also *reductionist.*

"Nonreductionist" Physicalism

A view of human agency and mind that came to great prominence within the last 40 years is often called "nonreductionist physicalism." In very general terms, such a view pares epistemological reductionism away from ontological reductionism, accepts only the latter but not the former, meaning that everything that exists consists of the objects, properties, relations, events, etc. of the physical world. Thus, the "physicalism" label. The "nonreductionism" label stems from two features of this view. First, the laws and theories of psychology and the other human sciences cannot

[21] For this usage, see Raphael von Riel and Robert Van Gulick, "Scientific Reduction," *Stanford Encyclopedia of Philosophy* (2014 revision), pp. 52–54. The label "epistemological" misleads here because the laws and theories that are said to be reduced in this sense are in no obvious way "epistemological." My own label for this kind of reduction has been, "nomological." (Moore, *Causation and Responsibility: An Essay in Law, Moral, and Metaphysics* (Oxford: Oxford University Press, 2009), p. 392. Robert Van Gulick's preferred label is, "representational." von Riel and Van Gulick, "Scientific Reduction," pp. 36–37.

be deduced from (or in some other way reduced to) the laws and theories of neuroscience. Second, the kinds of actions and mental states referred to in psychological laws and theories are not, as kinds, identical to certain physical kinds (such as bodies in motion for actions or events in the brain for mental states).

One might well wonder why such a view isn't either a kind of dualism or a kind of skeptical physicalism that we will shortly explore. As to the first of these possibilities: if the best explanations we have of human experience and behavior are irreducibly psychological, why doesn't adherence to such explanations commit one ontologically to the entities such theories are about, which are not (it is said) physical entities? Alternatively, if one takes the position that such psychological entities as intentions, beliefs, etc. don't really exist – because the only things that do exist are physical in nature and actions and mental states are not physical in nature because they cannot be reduced to bodily motions and brain states – then why are not all psychological/human science explanations simply false? In which case nonreductionist physicalism has all the appearance of pursuing a chimera, what David Armstrong once aptly called the "ontological free lunch."

The standard account of this would-be free lunch is in terms of four items: functionalism, multiple realizability, token-identity, and supervenience. Taking these in order, *functionalism* is the view that mental states (including those at the agential core of actions) are functionally specified states of the physical system that is the brain.[22] A functional state, as here used, is a state defined by its functional role, not its physical structure. Its functional role is that of a causal intermediary, being the effect of earlier events in the environment or the brain and being the cause of later events in the brain or in the environment. An intention, for example, is the kind of thing that it is by its ability to play two functional roles: first, its role in resolving the conflicts between conflicting sets of beliefs and desires, and second, its causally mediating role between such resolutions, on the one hand, and those actions done in response, on the other. Such roles are just physical relations between the physical structures each intention is posited to be.

There is thus no structural essence for a type of mental state such as an intention; there is only a "functional essence." Because such functions are functions *of* certain physical entities, events, or processes, functionalism

[22] See Moore, "Mind, Brain, and Unconscious," for an example of 1980s style functionalism about mind.

does not posit a kind of (dualistic) "functional reality." Still, a functionally defined mental state must be physically realized on each occasion of its existence. Yet such physical states, events, or processes as do realize a functionally defined type of mental state like an intention can vary from context to context. This is the *multiple realizability* thesis.[23] On one occasion, an intention to go downtown may be blip *X* in region *R* of the subject's brain; on another occasion such an intention may be blip *Y* in region *S* of the subject's brain. Still, all intention = tokens are *some* kind of process, etc., going on in the brain, even if a different process on each occasion of intention formation.

Mental states are thus said to be *token-identical* to brain states; each token of an intention is identical to some brain state token.[24] Denied, however, is any "type-identity," that is, any identification across the board of intentions-as-such with some discrete patch of goings-on in the brain.

Fourth and last, mental states are said to *supervene* on brain states, "Supervenience" is a relation imported into the philosophy of mind from meta-ethics.[25] There as here it means asymmetrical covariance: for every change of a moral property (or mental state) there must be some accompanying change of a natural property (or brain state), but not vice versa, that is, there can be changes of base-properties (natural properties or brain states, as the case may be) with no accompanying change in supervening properties (moral properties or mental states). This latter fact is but an implication of the multiple realizability thesis, for if mental states are multiply realized by different brain states, then a change from one realizing brain state to another realizing brain state will not necessarily be accompanied by any change in the (one and the same) mental state so realized.

My own view, like that of Jaegwon Kim,[26] is that the lunch being laid out by the "nonreductionist" physicalist is not as free ontologically as its

[23] Classically articulated in Hilary Putnam, "Psychological Predicates," in W. Caplan and D. Merrill, eds., *Art, Mind, and Religion* (Pittsburgh, PA: University of Pittsburgh Press, 1967); Jerry Fodor, "Special Sciences, or the Disunity of Science as a Working Hypothesis," *Synthese*, Vol. 28 (1974), pp. 97–115.

[24] Token-identity is briefly discussed in von Riel and Van Gulick, "Scientific Reduction," at pp. 29–30.

[25] For the use of supervenience in meta-ethics, see Michael Moore, "Moral Reality Revisited," in *Objectivity in Ethics and Law* (Aldershot, UK: Ashgate, 2004), pp. 193–201.

[26] Jaegwon Kim, "The Myth of Nonreductive Materialism," *Proceedings and Addresses of the American Philosophical Association*, Vol. 63 (1989), pp. 31–47; Kim, "The Very Idea of

proponents imagine. Start with multiple realizability: surely it will not prove to be the case that some mental state M can be realized by some distinct brain state B that differs on each occasion that M exists.[27] Surely, that is, M is realized by a finite disjunction of brain states, B_1 or B_2 or B_3 . . . or B_n, where n is not infinite and probably not even that large. If so, then M will be type-identical to the finite disjunction of brain states, even if it is not type-identical to any one disjunct.

Moreover, there must then be what Kim calls "local reductions": each disjunct of brain states (B_1, etc., in the earlier disjunction) will itself be type-identical to a class of brain state tokens. This, because given the finite nature of the disjunction, each disjunct will recur again and again as a realizer of M. (We should think that there will be such local reductions, even without this last consideration – for how would one ever know that some mental state token M_1 was (token) identical to some brain state token B_1 if that identity never recurred? Although one might (just barely!) be able to assign meaning to a truly token-only-identity, surely it is much more plausible to think that token M_1 is identical to token B_1 because type M is identical to type B.)

On this revisionist view of "nonreductionist" physicalism, the view becomes a kind of mind–brain reductionism. For on this revisionist view, every type of mental state M is identical to a type of brain state, namely, a type that consists of a disjunction of brain states, B_1 or B_2 . . . or B_n; and, in addition, each disjunct in the larger type-identity is itself a subtype of brain states that is realized by an indefinitely large number of actual and possible brain state tokens. So at bottom, even on this supposedly "non-reductionist" view, we are still "nothing but a pack of neurons," with all the challenges to moral agency that is thought to foster.

Skeptical Physicalisms: Eliminative Materialism

There is a skeptical version of a truly nonreductionist physicalism that merits its own separate treatment. This is the skepticism about the mind

Token Physicalism," in Simone Gozzano and Christopher Hill, eds., *New Perspectives on Type-Identity: The Mental and the Physical* (Cambridge: Cambridge University Press, 2012), pp. 167–185. See generally Kim, *Supervenience and Mind* (Cambridge: Cambridge University Press, 1993).

27 As some nonreductive physicalists seem to believe. See, e.g., Kim's former student, Terence Horgan, "Nonreductive Physicalism and the Explanatory Autonomy of Psychology," in S. Wagner and R. Warner, eds., *Naturalism: A Critical Appraisal* (South Bend, IN: University of Notre Dame Press, 1993), pp. 295–320.

that is, like all physicalisms, ontologically monistic about the world: what exists are only physical goings-on and the things necessary to best explain those physical goings-on (the latter claim to make allowances for scientific handmaidens like mathematics with its seemingly Platonist entities). This is also a skepticism about the mind that shares the "epistemological" view that mental state explanations, laws, and theories do not reduce to physicalist explanations, law, and theories. But unlike the more optimistic nonreductionist cousins examined in the previous subpart, these skeptics of the mind deny that psychological explanations compete successfully with physicalist explanations; rather, for various reasons (including the supposed impossibility of there being any brain states capable of exhibiting intentionality, for example), mental state explanations will lose out to a maturing science of the brain in explaining human behavior.[28] A successful neuroscience will thus *eliminate* (rather than find a place for) psychology and the mental states that that psychology presupposes. Thus the name for this kind of skepticism, "eliminative materialism."

It is an oddity in the history of ideas that eliminative materialism has come to be included as a kind of reductionist physicalism.[29] The oddity, because eliminative materialism is so plainly antireductionist and necessarily so to set up the explanatory competition in which psychology is supposed to be defeated by an advancing science of the brain. The inclusion nonetheless occurs because of the following fact about reductionist physicalism, a fact to which we have not yet adverted.

The post-Nagelian philosophy of science qualified Nagel's view of scientific reduction by observing that many reductions in science were not "smooth"; rather, they were "bumpy" (or "lumpy") in the sense that the science reduced had to undergo some change in its laws (and entities posited by those laws) for the reduction to go through.[30] Even temperature's reduction to mean kinetic energy, for example, was said to have a little lumpiness in it because of the need to stipulate a randomness of

[28] Paul Churchland, "Eliminative Materialism and the Propositional Attitudes," *Journal of Philosophy*, Vol. 78 (1981), pp. 67–90.

[29] The oddity is on display in full force in Michael Pardo and Dennis Patterson, *Minds, Brains, and Law: The Conceptual Foundations of Law and Neuroscience* (Oxford: Oxford University Press, 2013), where the authors assume not only that eliminative materialism is a form of reductionist physicalism (an assumptions shared by many others) but also that it is the only species of this genus.

[30] Kenneth Schaffner, "Approaches to Reduction," *Philosophy of Science*, Vol. 34 (1967); pp. 137–147; Clifford Hooker, "Toward a General Theory of Reduction," Parts I, II, III, *Dialogue*, Vol. 20 (1981), pp. 38–59, 201–346, 496–529.

direction in molecular motion to reach the identification needed. Lumpier still, it was thought, was the "mass" in Newton's theory in relation to the mass of which postrelativistic physics speaks; one has to work a bit on what mass means in Newtonian mechanics to get that to be but a special case of Einstein's theories. (The same for other concepts like "gravitation.")

At some point the force-fitting becomes too great to be plausible. The classic case is the refusal of Lavosier to identify his nearly discovered oxygen with the fire-explaining posit of the older theory of combustion, phlogiston. The chemical theory that preceded Lavosier treated phlogiston as a kind of fire retardant: the reason a lighted candle in a closed space eventually extinguished itself was said by the old theory to be because too much phlogiston had been created by the burning, and that to remedy the situation one had to "dephlogistinate" the air by opening a window so that the excess of phlogiston could escape. This is, of course, the opposite of what we would say about oxygen: it is the *absence* of oxygen, not its presence, that extinguishes the candle in a closed space, and it is the *presence* of oxygen, not its absence, that allows the candle to be reignited when the window is opened. Such disparities between oxygen and phlogiston, and between the laws that referred to them, were too great for Lavosier to identify phlogiston as really being (in lumpy form) oxygen; rather, Lavosier became a phlogiston-skeptic by denying both that it existed and that the laws that referred to it were true.

When theories and the entities to which they purport to refer are only "lumpy" in their differences (and thus still permit reduction), versus when those differences are so great that one theory must simply be judged false and the entities to which it purports to refer nonexistent, is a matter of degree. What is lumpily reducible shades into what is hostilely competitive. What would one say, for example, about pencils if some new theory showed us that they are really sentient spies from Mars – would we say that "it has turned out that pencils are spies from Mars" (reduction), or that "pencils do not exist" (skepticism)?[31]

Eliminative materialism thus shades into reductionist physicalism. Such shading explains why eliminative materialism is often classified as a kind of reductionist physicalism. But explaining a belief is one thing; justifying it is another. The Sorites Paradox shows us that the vagueness

[31] The thought experiment of Hilary Putnam in his "The Meaning of Meaning," in his *Mind, Language, and Reality* (Cambridge: Cambridge University Press, 1985). For what it is worth, Hilary's bet was that we would be reductionists about pencils.

of a continuum provides no reason to deny the existence of clear cases along that continuum. Some things are plainly heaps, some things are plainly not heaps, even if for other things it is unclear how they should be classified as to their heapiness. Because it matters to the kinds of moral skepticism one might have about responsibility whether one is a reductionist/physicalist or a skeptic of the mind (an eliminative materialist), I shall keep this skeptical form of physicalism distinct for the purposes of this chapter.

The Quietist Metaphysics of Componential Mechanism

Together with dualism, the three forms of physicalism we have examined constitute the four main players for plausible metaphysical pictures about the mind in relation to the brain. Of late, however, there have emerged two new entrants to this metaphysical tournament that claim to be resistant to the fourfold classification listed earlier. I describe the first of these here and the second in the next succeeding subpart.

In all fields of thought there are those who find the defence (or even the articulation) of any metaphysical position to be pointless, boring, or in some way unproductive. "Metaphysical" for them is a pejorative adjective. They thus remain quiet about their metaphysics. This is not so much a distinct metaphysical position as it is an attitude of distaste for the metaphysical enterprise.

In neuroscience and the philosophy of mind, philosophers such as Bill Bechtel[32] and Carl Craver[33] refuse to engage the positions described earlier. Rather, theirs is the more practical attitude of seeking an understanding of how the brain works to produce the mental life and the behavior that we all experience. Their views are explicitly mechanistic: human beings are machines ("mechanisms"), but they are fruitfully studied in the ways other complex machines are studied, in terms of subsystems within subsystems within subsystems . . ., etc., of mechanisms. This requires different "levels" of explanation, each level contributing to an overall explanation of what persons do and experience that is mechanistic through and through. Importantly, questions of reducing one level of explanation to another level do not arise for them; the only task

32 William Bechtel, *Mental Mechanisms: Philosophical Perspectives on Cognitive Neuroscience* (Mahwah, NJ: Erlbaum, 2007).

33 Carl Craver, *Explaining the Brain: Mechanisms and the Mosaic Unity of Neuroscience* (Oxford: Oxford University Press, 2007).

they find interesting is to explain how the mechanism works, being ecumenical about what kinds and levels of explanations can contribute.

I call such a view, "componential mechanism" – the label because the view is mechanistic and its strategy is to provide explanations in terms of the workings of ever finer-grained mechanical subsystems. I include the view as a variant of physicalism because the view has no truck with dualism. There are no ghosts posited or looked for in the components that make up the human machine. We are mechanisms all the way down.

The inclusion of componential mechanism as a kind of physicalism is thus apt for the purpose of this chapter. After all, componential mechanism holds the belief – that persons are mechanisms (machines) – that raises the challenge to responsibility that I here seek to isolate. Thus, even though the view is *quiet* as to precisely which kind of physicalistic metaphysics it might represent, the view speaks plainly enough to the issue relevant here. We are "nothing but a pack of neurons" on this view, no less than on the other forms of physicalism we have been examining.

The Sui Generis Metaphysics of "Grounding"

A number of contemporary philosophers such as Gideon Rosen,[34] Kit Fine,[35] Jonathan Schaffer,[36] and Selim Berker[37] (among others) share the impatience with traditional metaphysics of mind of the componential mechanists, yet they do not share the metaphysical abstemiousness. On the contrary, they boldly seek to articulate a new kind of physicalism, one that eschews the relations of deducibility and identity of the older reluctionists and that equally eschews the adequacy of functionalism (cum multiple realizability cum supervenience cum token-identity) of the "nonreductionist" physicalists. Rather, they posit the existence of a sui generis metaphysical relation of *grounding*. This unitary, primitive relation is supposed to do the work of identity in the older physicalisms, but

[34] Gideon Rosen, "Metaphysical Dependence: Grounding and Reduction," in B. Hale and A. Hoffmann, eds., *Modality, Metaphysics, Logic, and Epistemology* (Oxford: Oxford University Press, 2010), pp. 109–135.

[35] Jonathan Schaffer, "On What Grounds What," in D. Manley, D. Chalmers, and R. Wassermann, eds., *Metametaphysics: New Essays on the Foundations of Ontology* (Oxford: Oxford University Press, 2009), pp. 347–383.

[36] Kit Fine, "Guide to Ground," in F. Correia and B. Schneider, eds., *Metaphysical Grounding* (Cambridge: Cambridge University Press, 2012), pp. 37–80.

[37] Selim Berker, "The Unity of Grounding," working paper, 2015. I am indebted to Selim for educating me on the literature of grounding. Selim, one might say, gave me a good grounding in the literature.

better. On this view, one has mental state *M in virtue of* being in brain state *B*, but *M* is not identical with *B*. Rather, *B grounds M* by being a *more basic* fact than *M*. *M* does not ground *B* because grounding is an asymmetric relationship, unlike identity.

It is a matter of considerable dispute whether there is any sui generis relation of grounding, and if there is, whether this is the relation holding between mental states and brain states. Again, however, this is hardly a dispute we need to resolve here. For the "Young Turks" in metaphysics who defend grounding are, like the componential mechanists, physicalists through and through. For them, the more basic facts about the brain studied by neuroscience will ground the mental facts about the life of the mind, so that *at ground* – or, "basically" – we are nothing but the pack of neurons that Crick and others found to be such an astonishing hypothesis.

III The Salient Data of Neuroscience Making Some Form of Physicalism/Mechanism Plausible

It is the thesis of the immediately preceding part of this chapter that the distinctive challenge presented to responsible agency by physicalism does not vary between classical reductionist physicalism, so-called nonreductionist physicalism, componential mechanism, or the physicalism of those who *ground* mind in brain.[38] Each of these four forms of physicalism regard persons as machines, their actions and their mental states, as a machine's mechanical behavior and internal machine states. The idea that we are at bottom mere mechanisms is the insight that troubles many of those who view neuroscience as a threat to responsible agency.

These four metaphysical interpretations, however, are not to be confused with the actual data of neuroscience showing that we are but physical systems. Rather, such interpretations form the scaffolding into

[38] The omission of eliminative materialism from this list of physicalisms is deliberate. Eliminative materialism is omitted, not because it isn't a form of physicalism that casts us as mere mechanisms (machines), but because in addition to this challenge it issues its own distinctive challenge: there is no "we" to be a pack of neurons, there is only a pack of neurons. The skeptical challenge of eliminative materialism that is unique among physicalisms is the challenge that all we value about ourselves – our hopes, loves, goals, intentions, beliefs – don't exist, not in any separate realm of mind but not either in the respectably physicalist realm of the brain. The denial, in other words, is not that (e.g.) intentions are not brain states; it is that intentions are not anything. For these reasons, one must deal separately with eliminative materialism and its distinctive kind of challenge to responsible agency.

which neuroscientific conclusions might alternatively be put. Neuroscience's distinctive challenge to responsibility arises from the discoveries of neuroscience itself, not just from the metaphysical packaging in terms of which such discoveries are cast. In this part I wish to detail neuroscience's scientific case for its conclusion that we are, indeed, "nothing but a pack of neurons."

As was set out earlier, there are two relationships of greatest relevance in the present context. One is between human actions and certain physical goings-on, and the other is between states of mind and other sorts of physical goings-on. These two relationships are so salient here because they deal with the physical bases on which the two moral properties that determine overall moral blameworthiness and criminal liability are thought to depend: these moral properties are the properties of wrongdoing (breach of obligation) and culpability (breach of obligation in the wrongdoer's mind).[39] We shall examine the neuroscientific data about each of these metaphysical relationships in turn.

Actions as Brain-Caused Bodily Movements

With regard to human actions, before one gets to the neuroscience there is a bit of the philosophy of action that needs doing so we can isolate the "physical goings-on" with which actions are to be identified. Much has been written in the philosophy of action about just what this kind of thing human actions, in essence, might be. I limit serious candidates (there are others) for what human actions might be to these:

1. Bodily motions
2. Muscle movements
3. Other physical changes in the body
4. Willing (volition) of bodily motions

Some theorists have regarded one or other of these as being the essence of *all* human actions.[40] In this context, however, we can afford to be

[39] The details of how wrongdoing and culpability jointly constitute both moral blameworthiness and criminal liability are detailed in Michael Moore, *Placing Blame: A General Theory of the Criminal Law* (Oxford: Oxford University Press, 1997). A systematic overview is provided in Moore, "The Specialness of the General Part of the Criminal Law," in Dennis Baker, ed., *The Sanctity of Life and the Criminal Law: Essays in Honour of Glanville Williams* (Cambridge: Cambridge University Press, 2012), pp. 69–105.

[40] I discuss each class of such theorists in Moore, *Act and Crime: The Implications of the Philosophy of Action for Criminal Law* (Oxford: Oxford University Press, 1993, paperback

somewhat more eclectic. Most of the actions we do – like marrying another, hitting another, defaming another, etc. – we do by use of the one "Archimedean Lever" we have with which to change the world, our own bodies. Our finger moves, causing a trigger to move, causing a gun to fire, causing a person to be struck, causing that person to die. We have many descriptions of such an action in terms of the causal properties it possesses, descriptions like "killing," "shooting," "firing," "squeezing," and "flexing." Yet these are all descriptions of one and the same thing, the bodily movement of moving the finger.[41]

Sometimes, however, movement of our limbs is blocked; yet it is still an action of ours that one *tried* to move our finger. In such cases we might identify the action of trying to move the finger as a flexing of the relevant muscles and the relaxing of certain other muscles.[42]

With the advent of mind–brain interface machines we have new levers by which we can move the world. For example, a paralyzed actor wills the movement of his finger, knowing that he cannot move it but also knowing that if he tries to move it a slow negative shift in electrical potential will occur in the Supplementary Motor Area ("SMA") of his brain. If a mechanical reader of this shift is attached at the vortex of his skull above the SMA, it will detect the change in potential. If such a reader transmits a signal to the appropriate device, that device could turn on a light. The actor who knows how all this works and intends to use it to turn on the light, surely performs the action of turning on the light even though no movement of his body takes place.[43] The temptation is to identify his act as changing the electrical potential in his SMA, not a bodily motion as that phrase is normally understood, but a bodily process nonetheless.

Sometimes we do things – such as sum a group of numbers – completely in our head. There are such things as "mental actions,"[44] and

edit. 2010). The noneclectic view plumped for in the book is a combination of 1 and 4: acts are identified as willings-non-deviantly causing-bodily movements.

41 The view essentially of Donald Davidson, *Essays on Actions and Events* (Berkeley: University of California Press, 1980).

42 The view of Oliver Wendell Holmes, in his *The Common Law* (Boston: Little, Brown, 1881).

43 Argued for in Michael Moore, "Renewed Questions about the Causal Theory of Action," in J. H. Agular and A. A. Buckareff, eds., *Causing Human Action: New Perspectives on the Causal Theory of Action* (Cambridge, MA: MIT Press, 2010), pp. 27–43.

44 Peter Geach, *Mental Acts* (London: Routledge and Kegan-Paul, 1959).

these are actions unaccompanied by any bodily movements. Such acts are thus most readily identified as mental tryings, or willings.[45]

In assuming the relationship of neuroscience's data to human actions let us prescind from the debate between these three items and eclectically identify human actions as some combination of items 1–3 (reserving for later, separate discussion item 4). That is, on this view human actions are usually bodily movements (that are underlain by muscle movements themselves underlain by certain internal changes in the body), but some human actions are just muscle activities that do not produce bodily movements, and some are just physical changes in the body that neither are nor cause muscle activities or bodily movements.

Neuroscience has a well-developed, one-hundred-year-old story about these three physical items.[46] The story starts with the muscles that are the immediate cause of the movement of the limbs. The muscles are arranged into flexor/extensor pairs. Flexor muscles generally adduct or bring limbs toward the body, and extensor muscles generally abduct, or draw limbs away from the center of mass. Opponent flexor (e.g., biceps) and extensor (e.g., triceps) reflexes antagonize one another and promote opposite outcomes for the limb (flexion and extension, respectively). The basic neural circuits of these reflexes are independent and organized in parallel, and they have limited inputs and outputs, allowing for rapid, efficient processing. The cost of this efficiency, however, is that lower-level systems have limited integrative capacity. Moreover, they can be in conflict. Simultaneous stretch of both the flexor and extensor muscle may lead to a reflexive increase in muscle tension in both muscles, but because they are opposed in their actions, there may be no resultant limb movement.

Next up (in the motor system's causal genesis) from the muscles themselves are the motor end plates on the muscles. These plates are at

45 The "mental action" theorists identify all actions as such mental acts as tryings and willings. The locus classicus of the mental action theory is H. A. Prichard, *Moral Obligation* (Oxford: Oxford University Press, 1949), pp. 190–192. I discuss Prichard and more modern developments of the theory in Moore, *Act and Crime*, pp. 95–108.

46 The telling of the story below is a retelling of John Cacioppo and Michael Moore, "Decisions and Intentions Working Group Paper No. 1," MacArthur Foundation Project on Law and Neuroscience, 2008. I am indebted to my cochair of this research group, John Cacioppo, Director of the University of Chicago's Center for Cognitive and Social Neuroscience, for his doing the initial drafting of this working paper from which the discussion of the text is drawn. On this topic, see generally Susan Pockett, "The Neuroscience of Movement," in Pockett, Banks, and Gallagher, eds., *Does Consciousness Cause Behavior?* (Cambridge, MA: MIT Press, 2006).

the ends of axon fibrils and are attached to the muscles they stimulate. The neurons whose axon fibrils are attached to the muscles are of two types: efferent neurons, which transmit the signals to the end plates on both the extensor and flexor muscles and cause them to contract or to stretch, and afferent neurons, which transmit feedback information from the muscles back up the spine and allow for adjustments to be made, either by reflex or by higher, more complicated processes.

Next up the chain of the motor system is the spine and the rest of the "final" or "common" motor pathway. The muscles are innervated by two systems, the lower and the upper motoneuron systems. The lower motoneuron system, sometimes called the final common pathway, consists of cell bodies located in the spinal cord, brain stem, and pons with axon fibrils extending to motor end plates on muscle fibers. A lesion in any part of the lower motoneuron system results in a paralysis of the innervated effectors.

As documented by early investigators such as Sir Charles Sherrington, basic somatomotor control is effected at the level of the spinal cord, with spinal reflexes representing the lowest central level in somatomotor control systems. Spinal reflex circuits are relatively simple and may be comprised of a single central synapse. The monosynaptic stretch reflex, exemplified by the knee jerk reflex to the physician's rubber mallet, is one example. This reflex entails an afferent somatosensory link arising from muscle stretch receptors, which synapse directly on the lower motor neurons controlling that muscle. This simple circuit provides for a reflexive contraction of the stretched muscle, which tends to compensate for the perturbing stretch. Stretch reflexes exist in all major classes of somatic muscles.

Greater levels of integration in motor systems are achieved by hierarchical circuits that promote coordination among the basic spinal reflexes – what Sherrington referred to as the *alliance of reflexes*. In our flexor/extensor example, this entails a collateral projection of the stretch receptor afferents onto inhibitory interneuron circuit elements, which in turn project to and inhibit the motor neuron for the opposing muscle. Stretching the flexor muscle, for example, results not only in activation of the flexor motor neurons (stretch reflex), but also in inhibition of the opposing motor neurons via the inhibitory pathway. This exemplifies a general principle of neural organization articulated by Sherrington – the principle of *reciprocal innervation* – which stipulates that neural systems promote specific outcomes by activating the mechanisms for the target response while at the same time inhibiting opposing responses.

Paralleling the pattern for somatic motor control, adaptive and protective reflexes and mechanisms are organized at all levels of the neuraxis. The pain withdrawal reflex, for example, is organized at the level of the spinal cord and can be seen even after spinal transections that isolate the cord from higher brain systems. Pain withdrawal reflexes are protective reactions that arise from somatosensory afferents, carrying nociceptive signals, which anatomically link to flexor neuron pools by a multisynaptic spinal pathway. Through this spinal reflex circuit, noxious stimuli yield a protective flexor withdrawal response. Likely because of their adaptive value, pain withdrawal reflexes are among the earliest to develop and the most resistant to disruption. Although pain withdrawal reflexes at the level of the cord may not require the invocation of a construct of emotion or affect, they represent an important low-level evaluative mechanism for escape from noxious stimuli. Moreover, despite their neural simplicity, these circuits can show operant conditioning of escape, which can support not only escape from, but also active *avoidance* of, pain stimuli.

In contrast to the primitive avoidance system associated with flexor reflexes, separate mechanisms exist at the level of the cord for opposing extensor responses that promote engagement with the environmental stimuli. Reflexes such as the extensor reflex in response to nonpainful cutaneous stimulation of the palm or the sole of the foot contribute to postural, locomotory, and grasping responses that serve to engage the organism with the environment.

As considered earlier, flexor and extensor reflexes are organized largely in parallel, as they control distinct motoneuron pools for opposing muscles. Nevertheless, they do interact. These lower reflex substrates are integrated by higher-level circuits, such as those that implement reciprocal innervation, which tends to reduce concurrent activation. They are also impacted by even higher systems that contribute to volitional actions and confer a greater degree of flexibility and control over flexor and extensor motor neuron pools and lower reflex substrates. Thus, we can volitionally contract both flexor and extensor muscles (e.g., in stiffening the arm), which can overcome the lower-level reciprocal innervation, and we are able to override or suppress flexor pain withdrawal reflexes (e.g., to remove a sliver from the finger).

Further up in the causal chain producing voluntary motor movement is the second system mentioned earlier, the system consisting of the two upper motor pathways. Sherrington's alliance of reflexes does not stop with reciprocal innervation. The cell bodies of the lower motoneuron system are innervated by upper motoneurons from the pyramidal and extrapyramidal upper pathways. A lesion of each of these upper motoneuron pathways has distinctive behavioral effects. The cell bodies of the pyramidal upper motoneuron pathway are located in the precentral gyrus of the frontal lobe, also called the motor strip, and long axons project to the cell bodies of the lower motoneuron system in the pons, brain stem, and spinal cord. The pyramidal motor pathway controls all our consciously directed, deliberate movements. For instance, a lesion of the pyramidal tract from the facial region of the motor cortex to the cell bodies in the facial nerve nucleus results in the inability to self-consciously smile on the contralateral side of the face. If the afflicted individual hears a humorous joke, however, the spontaneity of the response produces a normal smile.

The ability exists to produce a symmetrical smile spontaneously/reflexively but not self-consciously because the extrapyramidal upper motoneuron pathway is still intact. The extrapyramidal upper motoneuron pathway is phylogenetically older than the pyramidal pathway and thus plays a relatively more important role in lower animals. The nuclei constituting the extrapyramidal system are manifold with the major parts located in the basal ganglia, including the caudate, putamen, and globus pallidus; the thalamus and subthalamic nucleus; and the substantia nigra and red nucleus in the midbrain. All of these nuclei are connected synaptically to one another, the brain stem, the cerebellum, and the pyramidal system. A lesion of the extrapyramidal innervation of the facial nerve nucleus leaves the afflicted individual capable of producing a willed or self-conscious smile, but the individual is incapable of producing a spontaneous smile when they hear something humorous.

As hierarchical levels are layered on the motor control system, progressively higher levels receive a wider array of inputs, have greater circuit complexity and computation capacity, and can achieve a broader and more flexible range of outputs. At the highest levels, the levels beyond the primary motor cortex, cerebral systems must process a tremendous amount of sensory information, integrate this information with associative networks, emotional/motivational substrates, and expectancies in the contexts of strategic goals and tactical plans. This requirement for enhanced information processing can impose a processing bottleneck that necessitates a slower, more serial mode of processing and selective attentional mechanisms. Although it is these highest-level systems that confer the greatest cognitive and behavioral capacity, they do not operate in isolation but depend on and interact with lower levels in the hierarchy.

Hierarchical dimensions of central nervous system organization can be demonstrated anatomically as well as functionally. The simple hierarchy, however, belies the true complexity of neurobehavioral substrates, as long ascending and descending pathways can bypass intermediate levels of hierarchical organization and interconnect across widely separated neural levels. Cortical motor neurons project not only to intermediate-level somatomotor networks, but also directly onto spinal motor neurons through long descending pathways. The long ascending and descending pathways in neural hierarchies, together with the existence of lateral interactions among elements (such as those that underlie reciprocal innervation), yield what has been termed a heterarchical organization. The outputs of a strict hierarchical system are coherent, as all levels are linked by intermediate regulatory levels, and all outputs are by the final common pathway. In a heterarchical organization, however, higher levels can directly access output mechanisms independent of intermediate levels. This organizational feature allows for concurrent expression of multiple rerepresentative systems, which can increase behavioral complexity but can also lead to functional conflicts. By self-consciously willing the stiffening of the leg, for example, higher-level systems can mask the stretch reflex and usurp control of motor neuron pools.

At the upper end of the pyramidal upper motoneuron pathway is the primary motor cortex. Separate areas of the primary motor cortex control separate body parts, such as tongue, lips, larynx, fingers, wrists, and hips. Unsurprisingly, the amount of primary motor cortex devoted to controlling a specific body part is proportional to the degree of complexity in the behavioral routines performed by that body part, not to the size of the part itself. When mapped in the 1930s by Wilder Penfield, it was thought that the individual muscles were controlled by the innervation of discrete neurons. Now the coding appears to be by movements of the various parts of the body, even though those movements involve coordinated innervation of neurons in the different areas of the primary motor cortex.

Anatomically next to the primary motor cortex is the Supplementary Motor Area and pre-Supplementary Motor Area ("pre-SMA"). These areas have long been known to be activated during the initiation of voluntary motor movement.[47] Of particular interest is their apparent role in choosing which action plan the motor system (primary motor cortex and below) will execute. A recent suggestion is that these areas serve one fundamental function: "The resolution of competition between motor plans so that a unitary action may emerge."[48]

Aha, one might think, here at last is where the roots of human agency are to be found in the brain. Yet neuroscience does not cash out those "roots" in terms of a *choosing agent*, or homunculus, for this would be to literally posit some primitively intelligent ghost who has its spiritual hands on the levers of the human machine. Rather:

> Positing a brain area that chooses the specific neural circuit required to execute the action leads to an infinite regress: one would have to explain what chooses its choice, and so on *ad infinitum* . . . a multitude of neural representations *must* be co-activated *automatically* by external stimuli, internal biases, and conditional associations between the two. There can be no homunculus determining which representation should be activated and which should not. Without a mechanism to resolve this competition, coherent action would fail to emerge because of interference from actions incompatible with the one being performed. Importantly, such a mechanism does not do any *selection* – as the competition itself is the selection – but merely reinforces the winners and inhibits the losers of the competition.[49]

Where one traces the causal ancestry of voluntary motor movement back past the SMA and pre-SMA systems is at the moment a matter of conjecture in neuroscience. The frontal regions have long been thought to be involved in executive functions such as formulating goals and plans, selecting among options to achieve these goals, monitoring the consequences of our actions in light of our goals, and inhibiting, switching and regulating our behaviors accordingly. Recent evidence indicates that the initiation of a motor response proceeds from the planning areas of the frontal cortex to the putamen, globus pallidus, thalamus, primary motor cortex, motor nucleus in the spinal cord, and finally to the muscles. Being able to inhibit a motor response once it has been initiated has obvious adaptive value, and this inhibition involves the right inferior frontal

[47] See the summary of early work here by Gary Goldberg, "Supplementary Motor Area Structure and Function: Review and Hypotheses," *Behavioral and Brain Sciences*, Vol. 8 (1985), pp. 567–588.

[48] Nachey, Wydell, O'Neil, Husain, and Kemard, "The Role of the Pre-Supplementary Motor Area in the Control of Actions," *Neuroimage*, Vol. 36 (2007), pp. T155–T163.

[49] Ibid. (emphasis in original).

cortex, which projects to the subthalamic nucleus (a region of the basal ganglia that may act on the globus pallidus to block the motor response). Monitoring for response conflicts, in turn, appears to involve the dorsal anterior cingulate and the adjacent pre-SMA motor area, which, in turn, is connected to the right inferior frontal cortex and subthalamic nucleus. Switching also involves the pre-SMA motor area and right inferior frontal cortex. Together, this work has led to the model discussed earlier, the model in which the pre-SMA motor area may monitor for conflict between an intended response and a countervailing signal, and when such conflict is detected the "brakes" could be put on via the connection between right inferior frontal cortex and the subthalamic nucleus region.

This completes my admittedly sketchy overview of the neuroscience of voluntary bodily movement in human beings. Yet even in the science so roughly sketched and still so much in development, surely it is overwhelmingly plausible to think that human action is at bottom a machine-like process. From the muscles themselves, the motor end plates on them, the axon fibrils of the lower or common motor pathway reaching to the end plates, the neurons in the two upper motor pathways, the body part and movement-specific activities of the primary motor cortex, the conflict-resolving and other mediating functions performed by the SMA and the pre-SMA, and on to the neuronal circuitry expressing goals and plans in the frontal regions of the brain, it is all rather plainly one piece of machinery – a very complicated and indeed elegant piece of machinery, to be sure, but still (just) a piece of machinery.

Mental States as Brain States

Unlike human actions, the mental states of persons do not come with an obvious "handle" in the physical furniture of the universe. There is nothing analogous to bodily movements or muscle activities for mental states that make the latter look on their face machine-like. Indeed, quite the contrary: on first blush mental states have a ghostly pallor, that of subjective experience and "intentional inexistence" of the objects of that experience.

Nonetheless, here too, neuroscience tells us, we are but machines. To get the flavor of this reduction of the mysterious to the mundane, consider this depiction of decision making by a fictional, only as yet imagined, neuroscience:

> At some time in the future we may have extremely high-resolution scanners that can simultaneously track the neural activity and

> connectivity of every neuron in a human brain, along with computers and software that can analyze and organize these data. Imagine, for example, watching a film of your brain choosing between soup and salad. The analysis software highlights the neurons pushing for soup in red and the neurons pushing for salad in blue. You zoom in and slow down the film, allowing yourself to trace the cause-and-effect relationship between individual neurons – the mind's clockwork revealed in arbitrary detail. You find the tipping-point moment at which the blue neurons in your prefrontal cortex out-fire the red neurons, seizing control of your pre-motor cortex and causing you to say, "I will have the salad, please."[50]

This is undoubtedly a simple-minded view of the mechanical process that underlies a decision about what to eat for lunch; still, no matter how much one techs up the neuroscience (so that decisions emerge from some much more complicated process than simple "outfiring" of pro-soup versus pro-salad neurons), Green and Cohen's bottom-line characterization will be accurate: "the mind's clockwork revealed in arbitrary detail."

Thus, although there is no obvious physical equipment on display with mental experience – no steam-shovel-like Archimedian levers (arms and legs), no obvious fulcrum points (joints), no power plants (muscles) – yet the reduction base for mental states has long been obvious. It is in the activity of the brain, what we now know to be a vast array of two valued switches organized in some complicated way that gives rise to mental experience.

Long before we knew very much about the basic mechanisms of the brain, it was obvious that mind depended on the brain in some very intimate way. Start where many psychology texts begin on this topic, with a railroad foreman of the nineteenth century named Phineas Gage.[51] Gage was tamping down an explosive charge on a railroad track in New Hampshire when he set off the charge by accident. This sent the tamping bar (a five-foot iron pike, basically) through his left eye socket and out the top of his skull, taking a good deal of his left prefrontal lobe with it. Remarkably, this did not kill Gage, who survived for another 11 years, nor even render him unconscious. Still, the loss of the brain tissue plainly changed his mental life dramatically and

[50] Joshua Greene and Jonathan Cohen, "For the Law Neuroscience Changes Nothing and Everything."

[51] John Fleischman, *Phineas Gage: A Gruesome but True Story About Brain Science* (New York: Houghton, Mifflin, 2002).

permanently, as described some years after treating Gage by his attending physician, John Martyn Harlow:

> The equilibrium or balance, so to speak, between his intellectual faculties and animal propensities, seems to have been destroyed. He is fitful, irreverent, indulging at times in the grossest profanity (which was not previously his custom), manifesting but little deference for his fellows, impatient of restraint or advice when it conflicts with his desires, at times pertinaciously obstinate, yet capricious and vacillating, devising many plans of future operations, which are no sooner arranged than they are abandoned in turn for others appearing more feasible. A child in his intellectual capacity and manifestations, he has the animal passions of a strong man. Previous to his injury, although untrained in the schools, he possessed a well-balanced mind, and was looked upon by those who knew him as shrewd, smart businessman, very energetic and persistent in executing all his plans of operation. In this regard his mind was radically changed, so decidedly that his friends and acquaintances said he was "no longer Gage."[52]

Even with the limited medical knowledge of 1848, the inference is well-nigh irresistible that Gage's mental life – his emotions, moods, capacity for commitment, desires, etc. – depended intimately on certain brain structures being present.[53]

Since 1848 neuroscientists have learned an enormous amount about the brain. About the two-valued decisions imagined by Green and Cohen, for example, John-Dylan Haynes and his collaborators have had some success in predicting the outcomes of such decisions some seven seconds or more prior to the subjects themselves knowing what they had decided,[54] from studying the oxygenated blood flow patterns in the precuneus region of the parietal cortex.[55] Likewise with respect to Gage's kinds of incapacities, the Damasios and their colleagues found

52 Ibid.

53 The striking similarities between the changed Gage and America's current president, Donald Trump, show that more than one constellation of brain structures can realize these mental characteristics (I am here making the generous assumption that Trump does have all of his prefrontal cortex). That is the lesson of the multiple realizability of mind by brain discussed earlier.

54 Or would decide – did the brain "decide" before the persons experienced it, or is this only a prediction-base for a decision not yet made? I raised this in my commentary on Haynes's presentation of his work at the Sage Center for the Study of the Human Mind, University of California, Santa Barbara, 2009, to which Haynes's response was that "that is for you philosophers to figure out."

55 John-Dylan Haynes, Katsuyuki Sakai, Geraint Rees, Sam Gilbert, Chris Frith, and Richard Passingham, "Reading Hidden Intentions in the Brain," *Current Biology*, Vol. 17 (2007), pp. 323–328; Chun Siang Soon, Marcel Brass, Hans-Jochen Heinze, and John-Dylan

that damage specifically to the ventromedial prefrontal cortex results in an insensitivity (a learning deficit) to the future consequences of persons' choices that they themselves highly value.[56] More recent work by Antonio Rangel and his colleagues further shows that modulation (of the signals headed from the ventromedial prefrontal cortex to the motor areas of the brain) must take place by signals originating in the dorsolateral prefrontal cortex; otherwise subjects show a marked incapacity to restrain their impulses (such as the impulse to eat unhealthy food while trying to diet).[57] Not surprisingly, given this dependence, damage to the dorsolateral prefrontal cortex results in an incapacity to restrain impulses, including, in one well-known case,[58] impulses of a sexual nature.

Such current neuroscientific knowledge does not yet understand the precise neural mechanisms that are responsible for these mental capacities and experiences, as much as such studies have done to isolate discrete areas of the brain where such mechanisms seem to be operating. We do know the basic physical equipment: the positive/negative ionization potential in neurons and the chemical transmission of that potential across synaptic clefts by the use of neurotransmitter vesicles. But what we don't yet know is how the organization of that basic physical equipment produces the capacities and experiences that we can measure behaviorally and phenomenologically. But surely acquiring that knowledge is just a matter of time.

The Metaphysical Interpretation of Current and Imagined Neuroscientific Data

When that time arrives, we will indeed see "the mind's clockwork revealed in arbitrary detail." By then we may also be able to settle to

Haynes, "Unconscious Determinants in Free Decisions in the Human Brain," *Nature Neuroscience*, Vol. 11 (2008), pp. 543–545.

56 Antoine Bechara, Antonio Damasio, Hanna Damasio, and Steven Anderson, "Insensitivity to Future Consequences Following Damage to Human Prefrontal Cortex," *Cognition*, Vol. 50 (1994), pp. 7–15.

57 Todd Hare, Colin Camerer, and Antonio Rangel, "Self-Control in Decision-Making Involves Modulation of the vmPFC System," *Science*, Vol. 324 (2009), pp. 646–648.

58 Jeffrey Burns and Russell Swerdlow, "Right Orbitofrontal Tumor with Pedophilia Symptom and Constructional Apraxis Sign," *Archives of General Neurology*, Vol. 60 (2003), p. 437. The neurologists in this case found that the subject's apparent inability to restrain expression of his sexual impulses was accompanied by a distortion of his dorsolateral prefrontal cortex by a large right orbitofrontal tumor. Ibid., at p. 128.

our satisfaction the correct metaphysical interpretation of the data. Is our restraining ourselves on a given occasion, for example: type identical to the mechanical workings within the dorsolateral prefrontal cortex? Token-identical, instantiating some "local" type-identity? Grounded in those mechanical workings? Best explained by this lower-level explanation in terms of mechanistic components? As we saw earlier in part II of this chapter, any of these metaphysical interpretations of the data will raise the challenge to responsibility that we seek to isolate and to judge.

The only metaphysical interpretations of the data that we definitely need to resolve for present purposes is whether the data supports either a dualistic metaphysics or the skeptical metaphysics of an eliminative materialism. For the first of these metaphysics would eliminate the challenge of physicalism because physicalism would then not be true; and the second of these metaphysics would present the distinct challenge to responsibility of eliminating morality's grounding in the natural properties of the folk psychology. In closing this part, I consider each of these possibilities in turn.

Dualism in Neuroscience?

The temptation toward dualism is not aided by any of the data of neuroscience, neither the data about actions nor the data about mental states. Rather, such an interpretation is forced onto the data of neuroscience by certain philosophical preconceptions, preconceptions such as that human agency and subjective experience *cannot* be analyzed in the nonagential, nonexperiential terms of physical science. These are philosophical conclusions, not deliverances forced on us by the data of neuroscience.

I personally find it surprising at this late date that so many of my fellow philosophers still bring these philosophical preconceptions of dualism to the table of science.[59] Admittedly, they feel driven to their dualistic conclusions by arguments they cannot answer, like the supposed "hard problem of consciousness," the supposed "new knowledge" gained when a zombie-like creature experiences color or sound for the first time,[60] the supposed "inexistence" of intentional objects, or the

[59] David Chalmers on his website keeps track of who is where on this issue, mourning the recent loss of Frank Jackson to dualism's cause but welcoming the (partial) subscription of Jaegwon Kim.

[60] The argument of Frank Jackson that he himself has repudiated but to which my longtime friend and (usual) comrade-in-arms, Richard Fumerton, has recently succumbed. See

difficulties of finding original intentionality in purely physical systems. This is not the place to engage the arguments of the new dualists in philosophy, although in truth I reject all of them.

But what can be done here is to issue a challenge to philosophical dualists: leave the philosophical pedestal and try working a metaphysical dualism into the explanations of good science. There is a famous exemplar of such an attempt: late in his career the Nobel Prize–winning Australian neurophysiologist Sir John Eccles turned from the experimental work to which he had devoted his professional life, to philosophizing (his terminology) about what must exist for us to be the persons and moral agents that Eccles rightly thought we manifestly are. He thus speculated about how will and the "psychons" that execute will, along with other dualistic mental things, could have a place within a scientifically respectable neuroscience. The results of such dualistic philosophizing by Eccles were not reassuring.

Of the three stumbling blocks keeping dualistic philosophers away from some form of monistic physicalism – qualia, intentionality, agency – Eccles was fixated on the last. He just couldn't see[61] how the passive brain events he did so much in his early work to reveal (like the chemical, not electrical, nature of synaptic crossings) could be adequate to explain the willings at the root of human agency. So he put willings front and center in his theory of human agency.[62] They were primitive, irreducibly mental things, things existing in time but not in space and things incapable of having, receiving, or transmitting energy.

How such immaterial things interact with the physical motor system that Eccles's teacher, Sherrington, did so much to detail is, of course, a stunning and obvious difficulty. He "solved" one aspect of the interaction problem (the conservation laws problem), he thought, by turning to quantum physics and the notion of a probability field. The will, he thought, worked by creating such probability fields to increase the probability of the movement willed occurring. Probability fields, he further posited, required no energy to be expended in either their creation or in their "influence" on the behavior whose probability is thus increased. Mental acts of willing could thus influence the

Richard Fumerton, *Knowledge, Thought, and the Case for Dualism* (Cambridge: Cambridge University Press, 2013).

61 Any more than could his teacher, Sir Charles Sherrington, who also explicitly adopted metaphysical dualism.

62 John Eccles, *How the Self Controls Its Brain* (Berlin: Springer Science, 1994).

Sherrington motor mechanism without violating the laws on the conservation of mass and energy.

He also dealt with another aspect of the interaction problem – does the interaction process itself have spatial location – by positing a "mind/brain interface line," the physical side of which was (at least early on) treated by Eccles as being located in the SMA.[63] If one takes Eccles's chart of this line literally, it seems that the line itself is inside the SMA – does this mean that the will too must be there, so that it too has spatial location? Hard to know, when one side of the line has definite physical location and the other is not supposed to have.

The mechanics of the interaction across the mind/brain interface line is also puzzling to the point of incoherence. Eccles speculated that on the physical side the neurons of the brain were (functionally, presumably, not structurally) bundled into 40 million packets each having about 100,000 synapses. The mind/brain interface worked by matching these 40 million "dendrons" (as he called these bundles) with 40 million "psychons." Psychons were conceived of as the micro mental events out of which a macro mental event like a willing was constituted. Yet unsolved, of course, by this miniaturizing of mind/brain interaction, is the mystery of the mechanics of the interaction. Positing 40 million dendron/psychon interactions does nothing to dispel the stupefying mystery surrounding just one such interaction.

I take it as settled that this explicit positing of a ghost in the machine is hopeless and is seen to be such by every serious student of mind and brain. Nonetheless, to be sure of this latter point, during a grand rounds-like tour of the various research departments of the John Curtin School of Medical Research on the campus of the Australian National University in Canberra (Eccles's longtime professional home), I asked each researcher whether they were looking for Eccles's psychons or even dendrons. With the exception of one (as it turned out, Buddhist) researcher, I found none. Indeed, I was politely but firmly told by the director to stop embarrassing them with such famously hopeless research designs. And in the larger world the results are even more unanimous: there is no way to put a primitively mental homunculus in the middle of the machine that is the brain and the body without giving rise to insurmountable problems.

[63] John C. Eccles, "The Initiation of Voluntary Movements by the Supplementary Motor Area," *Archives of Psychiatry and Neurological Sciences*, Vol. 231 (1982), pp. 423–439.

Eliminative Materialism

The other metaphysical response needing defusing here, skeptical physicalism ("eliminative materialism"), is not so decisively refutable. On the other hand, such skepticism is not that supported or supportable either. Eliminative materialism is mostly just a bit of speculation about what neuroscientific research might or could reveal. It is not an inference from the data so far produced and is thus far from being a thesis that we are compelled by the scientific evidence to accept. Will such research show us that nothing in the brain – not even a disjunction of multiply realizing states of the brain – correspond to the items of the folk psychology, items such as beliefs, desires, intentions, emotions, and the like? Without a more detailed understanding of the mechanisms that are at work when we, e.g., intend to go downtown now, it is difficult to say for sure. The bets of neuroscientific researchers all seem to be placed on the optimistic side of that question, however.[64]

There has been one example of a fully worked out eliminative materialism and that was the methodological behaviorism of B. F. Skinner.[65] Skinner famously claimed that we would be forced to abandon our conceptualizations of persons as autonomous, of them having either freedom or dignity.[66] We would be so forced by the success of Skinner's physicalistic scheme for explaining human behavior, a scheme that had no place for such concepts as intention and will in its explanations. Yet Skinner's version of skeptical physicalism was hostage to the success of his behavioral explanations in outcompeting the explanations of the folk psychology, a success his behavioral explanations never came close to achieving. True, the failure of Skinner's version of a skeptical physicalism

64 As are the bets of most philosophers on this. In assembling the invitees for the recent Roundtable on Metaphysics of Mind, Neuroscience, and Responsibility referenced in the swordnote earlier, my cohost Richard Fumerton and I had no luck finding any philosopher willing to come to defend eliminative materialism. Perhaps there should be no surprise there – Steven Stich gave up the position a long time ago, Dan Dennett was never properly so classified, and Paul and Patricia Churchland's own research (as opposed to what they sometimes said about it) belies any real commitment to the disappearance of BDI psychology.

65 As is well known, there are two kinds of Skinnerian behaviorism: the methodological behaviorism discussed in the text, according to which explanations framed in the concepts of the folk psychology will lose out to those framed in the language of stimulus and response, and the logical behaviorism that Skinner also (and inconsistently) advocated, according to which folk psychological terms like "intentions" could be *reduced* to environmental reinforcers and dispositions to behavior. See Daniel Dennett's "Skinner Skinned" in his *Brainstorms* (Putney, VT: Bradford Books, 1978) for the difference.

66 Most popularly, in B. F. Skinner, *Beyond Freedom and Dignity* (New York: Knopf, 1973).

cannot be a guarantee of a similar failure of such a neuroscience-based physicalism where the science is better. Even so, nothing that better science has yet produced should incline us to reject the folk psychology on which our notions of moral responsibility are built.

IV The Implications of the Astonishing Hypothesis on Moral Responsibility

Framing the Question

Why do people think that moral responsibility and blameworthiness – and the institution of the criminal law insofar as it is built on these moral notions – are threatened by the promise of neuroscience to show that physicalism (in one of the four forms distinguished in part II of this chapter) is true? This requires a bit of a delicate dance for me. This, because my bottom line is that people should not at the end of the day believe that responsibility *is* threatened by physicalism. Yet to answer the question asked here, I need to tease out why one might think it is. I need views plausible enough to be worth considering, while not so plausible as to be true.

Two classes of arguments can be put aside at the start. The first such class consists of arguments drawing conclusions skeptical of responsibility from *implications* of physicalism rather than from physicalism itself. I refer to arguments: (1) showing that physicalism implies determinism, and inferring that no one is responsible if determinism is true; (2) showing that physicalism entails epiphenomenalism of choices and willings vis-à-vis the actions they potentially cause, and inferring that no one is responsible if the will is only epiphenomenal; or (3) showing that physical systems can have no privileged access to their internal states and inferring that without the inference-less and certain knowledge a person has of his own mental states, no one can be responsible.

These three modes of argument do indeed proceed to skepticism about responsibility from physicalism about mind and action. Yet such arguments simply duplicate the separate arguments from determinism, epiphenomenalism, and fallibilism that I put aside at the start. Some distinctive argument is wanted here for skepticism about responsibility proceeding from physicalism alone, unaided by arguments from these other three sources.[67]

[67] I thus put aside those like Adam Kolber, whose skepticism about responsibility just cannot separate physicalism proper and physicalism's deterministic corollary. E.g., Adam

The second class of arguments to be put aside are arguments that proceed through eliminative materialism. These are the arguments that rehearse the familiar objections to mind–brain reduction in any of the four forms that I distinguished in part II of this chapter. Those objections were that (1) *qualia* (mental experience) cannot be found in the physical world; (2) physical stuff is not *about* other physical stuff so that physical systems cannot exhibit intentionality; and (3) agency is not to be found in the merely passive states of affairs and the causal relations between them that populate the physical world. From this triad of objections, the conclusion is drawn that the entities presupposed in the folk psychology (principally the belief, desires, intentions that make up "BDI" psychology) do not exist, and concluding therefore that moral responsibility cannot exist either.

This second form of argument productive of skepticism about responsibility does proceed from physicalism, yet it is not the kind of physicalism on which I wish to focus. Sure enough, eliminative materialism about the mental will produce skepticism about the moral.[68] If it were true. But as was set forth at the close of part III, there is no reason to believe that eliminative materialism is true. I thus focus here on the four kinds of physicalism that neuroscientific research holds out the promise of showing to be true. My question is: Does *that kind* of physicalism have any tendency to make us skeptical about moral responsibility? Later I discuss my three best guesses about this.

Moral Responsibility Requires Dualism to Be True, and Physicalism Shows Dualism to Be False

This basis for skepticism about responsibility just barely crosses the threshold of plausibility that makes it worth even discussing. It is not that physicalism in all its forms does not rule out dualism – it does.

Kolber, "Will There Be a Neurolaw Revolution?" *Indiana Law Journal*, Vol. 85 (2014), pp. 807–845:

> [Y]ou are not responsible for the billions of sometimes randomly colliding particles that make you ... [do what you do]. How can we be responsible for actions that follow mechanically from forces beyond our control? ... If we are mechanisms – like clocks that tick but have no moral agency – we arguably cannot be responsible at all.

68 Eliminative materialism would be anathema for responsibility if it were true. E.g., Jerry Fodor, *Psychosemantics* (Cambridge, MA: MIT Press, 1987):

> If common sense psychology were really to collapse, that would be beyond comparison the greatest intellectual catastrophe in the history of the species.

The difficulty rather is finding any reason to think that moral responsibility requires dualism.

The most popular routes to believing that we cannot be responsible without dualism being true are blocked by the way I have framed the question. For example, as Gilbert Ryle charted long ago,[69] many people accept the incompatibilist idea that no one can be responsible unless their choices and their actions are free ("free" in the sense of "uncaused"). Then, the inference is, for such magical freedom from determinism to exist, it needs a magical kingdom in which to reside, one immune to the forces of the physical world: voila, dualism! As another example, many other people (of somewhat greater sophistication) think that responsibility requires that mental states like intentions must possess the causal efficacy that folk psychology says they have, i.e., that the epiphenomenalism of the will must be false. Then, the inference is, for there to be a distinctly mental causation dualism about the mental must be true – otherwise the putative mental causes are really just physical causes.[70]

What is wanted here are arguments for why or how responsibility presupposes the truth of dualism different from the popular arguments given earlier. For arguments like those earlier described duplicate the well-known, free-standing arguments for dualism that proceed from finding the negation of determinism or epiphenomenalism to be plausible. Yet as I set the question that this essay seeks to answer, what is wanted are arguments for moral skepticism that proceed from physicalism/reductionism alone, unaided by the affirmation or the denial of determinism, epiphenomenalism, or falliblism.

Such arguments are hard to find in the literature. Two legal theorists, Dov Fox and Alex Stein, have recently sought to show how dualism is presupposed in those of our responsibility assessments as are reflected in Anglo-American criminal law.[71] Fox and Stein find dualistic

[69] Gilbert Ryle, *The Concept of Mind* (London: Hutcheson, 1949). It has long been known that positing the existence of a magical kingdom where the will is immune to causal influence is hard to square with the will's need to have causal influence outside the kingdom.

[70] The voluminous literature on this is assessed in Brian Leiter and Alexander Miller, "Closet Dualism and Mental Causation," *Canadian Journal of Philosophy*, Vol. 28 (1998), pp. 161–181. As Leiter and Miller point out, the need for there to be mental causation (rather than epiphenomenalism of the will) is as easily satisfied by a reductionist physicalism as by any dualism.

[71] Dov Fox and Alex Stein, "Dualism and Doctrine," in Dennis Patterson and Michael Pardo, *Philosophical Foundations of Law and Neuroscience* (Oxford: Oxford University Press, 2016), at pp. 126–135.

suppositions in the reliance on intentions by both morality and the criminal law. Although they at the start admit that "[d]ualism's presence in the criminal law's concept of intentionality is less [than] straightforward,"[72] they nonetheless think that the way intention is used to ascribe responsibility is dualistic.

The basic idea goes like this. Criminal law separates actus reus from mens rea, both being required for prima facie criminal liability. Analogously, morality separates wrongdoing from culpability, both being required for prima facie blameworthiness and responsibility. Such separation of these two requirements, it is said, reifies mental states of mens rea like intention as if it could exist without the action intended occurring. Apparently it is metaphysically dualist to do this, they say, because "mental states are embedded in and realized by the conduct they drive."[73] As they elsewhere put it, metaphysical dualism is presupposed by any approach to mens rea whereby "a person's intent has a subjective status independent of whether his actions manifest the intent as an objective matter."[74] Apparently a "person's words and actions" must "constitute criminal intent," and not merely "serve as evidence of it," in order not to be metaphysically dualist;[75] or again, by conceiving of intention a something that need not be "embedded in his action . . . this . . . reinforces the dualist platform of subjective culpability and mindifies the doctrine of intentionality."[76]

This argument by Fox and Stein (and, behind them, by Fletcher) is a giant non sequitur. Nothing is presupposed about dualism by rejecting the idea that intention is necessarily to be found ("constituted") in action. *All* that follows from this rejection is that intentions are not actions and that they can exist independent of their execution in action. (A rejection, incidentally, that seems manifestly true.) *Nothing* is said or implied about whether intentions are or are not identical to, grounded in, brain states. Nothing.

[72] Ibid., p. 169. [73] Ibid., p. 129 [74] Ibid., p. 128. [75] Ibid.

[76] Ibid., p. 139. For this last proposition Fox and Stein quote George Fletcher. See if this helps:

> This pattern of liability presupposes a notion of intending that treats intent as a dimension of experience totally distinct from external behavior. Intending is conceived as an event of consciousness, known to the person with the intent but not to others. Thus, the relationship of intending to action is dualistic rather than monistic. The intent exists in the realm of the mind, the act in the realm of the body. (George Fletcher, *Rethinking Criminal Law* (Boston: Little, Brown 1978, pp. 118–119)

"*Thus*?" As the saying goes, I get everything in Fletcher's argument chain here but the "thus."

This Fox/Stein/Fletcher argument is thus a nonstarter. Still – looking for arguments on how physicalism leads to skepticism, this one was plausible enough to draw in at least three theorists of responsibility. But manifestly, whatever its level of plausibility, this basis for skepticism that proceeds from physicalism is ill-conceived. Moral responsibility does not presuppose dualism, so knocking down that supposed presupposition cannot explain the tendency to infer skepticism about moral responsibility from physicalism.[77]

Reduction as Demotion: Reduced Properties and Entities as Second-Class Citizens in Ontology's Republic

All four forms of physicalism here pertinent are in some sense reductionist, as we saw in part II of this chapter. For some people it seems that reductionism itself promotes skepticism about responsibility because it casts a pall of fictionalism about the mental states and actions on which responsibility depends. On this view of reduction, to be reduced is to be demoted to a second-class, almost fictional, status in the ontology of what (really) exists. First-class status is awarded only to the base properties and entities to which reductions are made, not to the items reduced to them. The "nothing but a pack of neurons" quotation from Crick with which we started has this flavor to it. Minds, emotions, hopes, etc., are actually "nothing but" patterns of neural discharge, implying that what really exists are only the physical energy transfers.

To the extent that reductions rely on *identifying* minds and actions with physical states and processes, this is a problematic basis for skepticism. After all, identity is a two-way street: if water is identical to H_2O, or temperature of a gas is identical to the mean kinetic energy of its molecules in random motion, then it is as true that water and temperature exist as that H_2O and mean kinetic energy exist.

Yet it is this very symmetry of identity that raises problems for it as the touchstone of ontological reduction. Reduction is thought to be an asymmetrical relationship, so that water reduces to H_2O, but *not* the

[77] If Fox and Stein are right that their efforts here are "the first to investigate … the prominence of mind-body dualism in our legal system," and that "We provide that systematic analysis for the first time here." (ibid., p. 107), that would be another reason to have included their argument here, despite its manifest weakness.

other way around. So if reduction is analyzed in terms of identity, that relation must be supplemented by some other criterion of reduction that introduces some asymmetrical relation of "more basic than." (Alternatively, of course, one leaves identity entirely, and analyzes ontological reduction in terms of some primitive grounding relation.)

Thus the skeptical worry recurs: if neuronal activity is more basic than the mental activity that reduces to it, is the mental activity in any way inferior in its ontological status? I don't see why. No matter how one cashes out the asymmetrical relationship between water and H_2O, or between temperature and kinetic energy, there really is water and there really is temperature, despite their reducibility. Analogously, there really are intentions (and thus there can be the responsibility that depends on there being intentions) even if intentions are identical to, functionally realized in, or grounded on, states of the brain.

Moreover, if anything like the story told by so-called nonreductive physicalism is true, the reduced entities like intentions may have pride of place when it comes to explanatory fecundity. One of the hypotheses of this metaphysics of mind is that the most salient causal *laws* (in terms of which good explanations are framed) are to be found at the level of the reduced entities; not at the level of the reducing entities. E.g., the ideal gas laws are best framed in terms of the macro properties of temperature, pressure, and volume, not in terms of the velocities and directions of individual molecules determinative of kinetic energy. To the extent that existence is a matter of having a necessary place in sound explanations, the ontologically reduced entities may not be "epistemologically" second-class citizens at all.

Paul Grice once made a virtue out of taking seriously whatever entities are needed to make our best explanations true:

> I am not greatly enamored of ... a concern to exclude such "queer" or "mysterious" entities as souls, purely mental events, purely mental properties, and so forth. My taste is for keeping open house for all sorts of conditions or entities, just so long as when they come in they help with the housework. Provided that I see them work ... I do not find them queer or mysterious at all ... To exclude honest working entities seems to me like metaphysical snobbery, a reluctance to be seen in the company of any but the best objects.[78]

[78] H. P. Grice, "Method in Philosophical Psychology (From the Banal to the Bizarre)," *Proceedings and Addresses of the American Philosophical Assoc.*, Vol. 48 (1975), pp. 23–53, at 30–31.

The Reductions of Mind to Brain, and Actions to Bodily Processes, Are "Skeptical Reductions"

Existing in the twilight zone between skeptical physicalism ("eliminative materialism") and lumpy version of classically reductionist physicalism is what Saul Kripke once called "skeptical reductions."[79] The skepticism of a skeptical reduction is not a full-blown skepticism; the skeptical reductionist does not deny the existence of the item reduced, as does a true (or "full-blown") skeptic. Yet what marks the reduction as skeptical is that the base item to which the reduction is made is so different than one expected that a large part of one's concept or stereotype of what the item reduced was really like, has to be sacrificed.

Examples do better than analysis here. Kripke's example is Hume on causation. Famously, Hume reduced the causal relation to the regular succession of spatio/temporally close classes of events.[80] He did this in two steps: first, he reduced singular causal *relations* to instantiations (at a certain time and place) of causal *laws*; and second, he reduced causal laws to the regular succession of spatio/temporally close classes of events. Kripke classified Hume's reduction as a skeptical one because it leaves out of causation what many, pre-Hume, thought essential to it, viz., the "glue" that links one event to another by a kind of "making happen" or "necessitation" relation.

Other examples of skeptical reductions are familiar in philosophy and in science, Skinner's *logical* (versus methodological) behaviorism was based on the skeptical reduction of mental states to behavior, either actual or hypothetical. This isn't classed as an outright skepticism because the existence of mental states is not denied; yet it is skeptical in that on this view of them mental states like intentions have no causal power over the actions that are their objects, contrary to what many thought to be essential to such mental states.

One of the motives for skeptical reductions can be glimpsed from these two examples. This is the motive of "solving" (really, avoiding) the "gap" problem, the problem that often there is a gap between the evidence available to verify the presence of something and the thing the evidence is evidence of. The verificationist "solution" is to identify the two so as to eliminate the gap. That events of class *X* are regularly succeeded by events of class *Y* may be the evidence by which we verify

[79] Kripke's actual phrase was "skeptical solutions." Saul Kripke, *Wittgenstein on Rules and Private Language* (Cambridge, MA: Harvard University Press, 1982), pp. 66–68.

[80] Described in Moore, *Causation and Responsibility*, p. 90.

the presence of a causal law between *X* and *Y*. Yet the *X*/*Y* causal law seems to say more; it seems to say that all instances of *X* will be followed by instances of *Y*, that being *X necessitates* being *Y*. There is a gap between evidence and the thing evidenced. The skeptical solution is to identify causal laws with regular succession, for then there is no gap between the thing evidenced and the evidence for it. (*Mutatis mutandis* for behavioral evidence, and the mental states evidenced by that behavior.)[81]

Sometimes skeptical reductions are motivated by considerations other than verificationist fears of evidential gaps. Often this motive is the ontological embarrassment Paul Grice mentioned at the close of the last subpart: if an entity seems queer, strange, or mysterious, reduce it to a nonqueer, familiar, nonmysterious entity whose existence is unproblematic. Although some of this fear of queerness motivated both Hume and Skinner to their skeptical reductions, consider a third example. Some moral theorists wish to identify moral properties with the moral beliefs of historical persons, either individually ("subjectivist meta-ethics") or collectively ("relativistic meta-ethics"). As Geoff Sayre-McCord recognizes,[82] in some sense these subjectivists and relativists are realists about moral properties, for they do affirm the existence of such things. Yet the subjectivist and relativist reductions are skeptical ones because they leave out what many would have thought crucial to metaphysical realism about moral properties, namely, their "mind-independence" in a robust sense of that familiar phrase. Surely many who make these skeptical reductions are motivated to do so by their embarrassment at having no other way that they see to include moral properties in their ontology.

Returning to the specific topic at hand, why would one think that the reductions of mind and action to physical things (in the four senses of reduction earlier set forth) would be a *skeptical* reduction? What is left out by such reductions that seem so crucial to our pretheoretical conception of our own minds, our deeds, ourselves?

[81] This verificationist motive is also clearly on display in Nelson Goodman's skeptical reduction of when a predicate is projectable to when a predicate has been, in the past, projected. Such reduction was supposed to solve what Goodman called the "new riddle of induction," but of course it solves that riddle only in the Pickwickian sense of a skeptical solution. See Nelson Goodman, *Fact, Fiction, and Forecast*, 4th ed. (Indianapolis, IN: Bobbs-Merrill, 1984).

[82] Geoffrey Sayre-McCord, "The Many Moral Realisms," in G. Sayre-McCord, ed., *Essays on Moral Realism* (Ithaca, NY: Cornell University Press, 1988).

Well, not subjective experience ("qualia"), intentionality ("meaning"), or agency – for the reductionisms we are here assuming to be true can account for all of those (if they couldn't, these reductionisms would not be true). Not freedom of action and will, causal efficacy of will, or privileged access to the contents of our own minds – for these one must in this context assume we can have even if we are machines (because we started by putting aside these separate bases for skepticism). Not the absence of any immortal soul-stuff in the reductionist picture – maybe some religious persons have this at the center of what responsible agents have to have to be responsible agents, but for us atheistic moral realists, no such postulate is part of our sense of who we are or have to be for us to be responsible.

What then? Is it that there is no room in the reductionist view of persons to allow for the existence of a homunculus within us who sits at the edge of some precipice of a synaptic cleft throwing vesicles into the void, or who sits in the control room of the human steam-shovel pulling all the levers that make it go? We can be disappointed by the absence of such an executive homunculus only if we were silly enough to expect that there is a little person inside us who is just like us in all of our qualities save she/he is much smaller. And surely the looming regress in homuncular explanations is too apparent for the absence of such creatures to shock us.

What the temptation to homuncularism shows is perhaps not as silly as homuncularism itself. What the temptation shows is how holistic we are in how we view ourselves, holistic in the sense that we think of ourselves as a unified whole and not as subsystems within subsystems within subsystems . . . etc. Our experience is of unity (most of the time, anyway). Yet the reality that reductionism teaches is one of physically realized functional subdivisions giving us many components and parts. This does contradict something basic in our experience of ourselves.

Yet is this any more than a kind of surprise, of a piece with those that science springs on us all the time, about everything? We first experienced heat as a sensation; isn't it amazing that what we were experiencing was really a bunch of (nearly) perfectly elastic billiard balls bounding into each other? Dan Dennett nicely captures how surprising science – including brain science – can be:

> In fact, of course, science is an unparalleled playground of the imagination, populated by unlikely characters with wonderful names (messenger RNA, black holes, quarks) and capable of performing the most amazing

> deeds: sub-atomic whirling dervishes that can be in several places – everywhere and nowhere – at the same time; molecular hoop-snakes biting their own tails; self-copying spiral staircases bearing coded instructions; miniature keys searching for the locks in which they fit, on floating odysseys in a trillion synaptic gulfs.[83]

Simple surprise, even "astonishment," cannot mark a reduction as a skeptical one.

I thus find it a genuine puzzle what it is people think they lose if they were to accept the reductionist picture of mind and action. Given the incredulity or even ferocity with which such reductionist theses are sometimes met, I gather it is something. Just not something that matters very much, not really. In 2003 Dan Dennett gave an interview that was headlined the next day as, "Yes we have a soul but it's made of lots of tiny robots." As Dennett afterwards commented on this headline:

> That's exactly right. Yes we have a soul, but it's mechanical. But it's still a soul, it still does the work that a soul was supposed to do. It is the seat of reason. It is the seat of moral responsibility. It is why we are appropriate objects of punishment when we do evil things, why we deserve the praise when we do good things. It's just not a mysterious lump of wonder stuff that will outlive us.[84]

Just so.

V Conclusion

So – is there a skepticism about responsibility that uniquely originates in reductionism/mechanism/physicalism and is not dependent on determinism, epiphenomenalism, or fallibilism? Certainly not in the sense that there is some inference from reductionism that should *truly* make us skeptical of responsibility. But then none of the other neuroscience-based skepticisms about responsibility that I have examined elsewhere (determinism, epiphenomenalism, fallibilism) are *true* either. The hypothesis that motivated the writing of this chapter was less ambitious: it only sought to separate an inference of skepticism from physicalism (from like inferences originating instead in determinism, epiphenomenalism, or fallibilism) that was *plausible enough* to be tempting. In truth, I am not

[83] "Reflections on 'A Conversation with Einstein's Brain'" in *The Mind's I* (1981), edited by Douglas R. Hofstadter and Daniel C. Dennett.

[84] "Atheism Tapes, part 6", BBC TV documentation of Jonathan Miller, produced by Richard Denton, recorded 2003, broadcast 2004.

sure that I have succeeded in even that more modest task. I myself find physicalism to be sufficiently nonthreatening to responsibility that I have difficulty gauging how threatening others find it to be.[85]

[85] I have some limited data here. Just before I was to give a version of this chapter at the University of Pennsylvania, Center for Neuroscience and Society, my wife and usual philosophical collaborator, Heidi Hurd, e-mailed me her views on the separation of the determinist skeptic from the reductionist skeptic:

> Just say, "there ARE two different debates here about TWO different things that persons standardly conflate" very loudly and repetitively (with that impatient, slightly disdainful tone in your voice that you always get when you are on philosophically thin ice), and people will be sure that they're the idiots when they can't see why the distinction you are drawing makes any difference!!

3

Non-Eliminative Reductionism: Not the Theory of Mind Some Responsibility Theorists Want, but the One They Need

KATRINA SIFFERD*

In this chapter I will argue that the criminal law is most compatible with a specific theory regarding the mind–body relationship – non-eliminative reductionism. Criminal responsibility rests upon mental causation: a defendant is found criminally responsible for an act where she possesses certain culpable mental states that are causally related to criminal harm in the appropriate way. If we assume the widely accepted positon of ontological physicalism, which holds that only one sort of thing exists in the world – physical stuff – non-eliminative reductionism about the mind offers the most plausible account of the full-fledged mental causation criminal responsibility requires. Other theories of the mind–body relationship, including eliminativism and non-reductive physicalism, threaten criminal responsibility because they do not offer satisfactory accounts of mental causation. Eliminativism, as the name implies, eliminates the mental or is skeptical that it can do the causal work necessary to responsibility; and non-reductive theories disconnect the mental from the physical/casual world such that the mental can no longer have reliable causal effects.

Mental Causation and the Criminal Law

To be criminally responsible, a defendant must be found to possess certain mental states at the time a crime was committed, and these mental states must be causally related to criminal harm for which the

* The author would like to thank Ty Fagan and Bill Hirstein for their very helpful comments on an earlier draft of this chapter, and Michael S. Moore and Stephen Morse for inspiring and critically engaging with the ideas presented here. Thank you also to David Papineau, who taught me to be a physicalist.

defendant has been arrested. This means that the criminal law places mental states in a privileged role in the explanation of human action, where such states are seen as the source or cause of behavior.[1] The theory of human psychology grounding the criminal law is a folk, or commonsense, psychological theory.[2] This means the criminal law trades in, and has codified, the psychological theory that most human beings naturally utilize to attribute psychological states to themselves and others when they attempt to explain or understand human behavior.[3] The folk theory of psychology, also called our "theory of mind," emerges very early in human development and uses behavioral cues, as well as knowledge about persons' character and preferences, etc., to attribute mental states to others[4]. So, even fairly young children pass the "false belief" test, which is a measure of whether a subject can attribute specific belief states that they themselves do not hold: a child as young as three can figure out that a doll who last viewed a toy in a chest will look for it in the chest, even though the test administrator moved the toy out of the chest when the doll was "away."[5]

1 On this, see: Morse, S. "Neuroscience and the Future of Personhood and Responsibility" in J. W. Rosen ed. *Constitution 3.0: Freedom and Technological Change* (Washington, DC: Brookings, 2011); Morse, S. "Inevitable Mens Rea" (2003) 27 *Harvard Journal of Law & Public Policy*; Morse, S. "Criminal Responsibility and the Disappearing Person" (2007) 28 *Cardozo Law Review*, 2545–2575; and Sifferd, K. L. "In Defense of the Use of Commonsense Psychology in the Criminal Law" (2006) 25 *Law and Philosophy*, 571–612. Prominent legal scholar Stephen Morse claims that our ordinary understanding of human behavior posits that "virtually all actions for which agents deserve to be praised, blamed, rewarded, or punished are the product of mental causation" (Morse, ibid, 2011, at 530).

2 See Sifferd, ibid.

3 One group of persons who may not have this commonsense "theory of mind" necessary to attribute mental states to others is autistics. See Baron-Cohen, S. *Mindblindness: An Essay on Autism and Theory of Mind: Learning, Development and Conceptual Change* (Cambridge, MA: MIT Press, 1995). For more information on folk psychology, see: Fodor, J. *Psychosemantics: The Problem of Meaning in the Philosophy of Mind* (Cambridge, MA: MIT Press, 1987); Jackson, F. and P. Pettit "In Defense of Folk Psychology" (1990) 59 *Philosophical Studies*, 31–54; Morse, S. "Determinism and the Death of Folk Psychology: Two Challenges to Responsibility from Neuroscience." (2008) 9(1) *Minnesota Journal of Law, Science and Technology*, 1–36.

4 Gopnick, A. and H. M. Wellman, "Why the Child's Theory of Mind Really Is a Theory" (1992) 7 *Mind & Language*, 145–171.

5 Wimmer, H. and J. Permer, "Beliefs about Beliefs: Representation and Constraining Function of Wrong Beliefs in Young Children's Understanding of Deception" (1983) 13 *Cognition*, 103–128.

The U.S. Model Penal Code asks courts to sort the folk psychological mental states assigned to criminal defendants into four defined categories: (1) purposeful, (2) knowing, (3) reckless, and (4) negligent.[6] For example, a homicide might be committed where the death was caused "purposely" (e.g., where the shot was fired for the purpose of killing the shooter's correctly identified target); "knowingly" (e.g., where the shooter actually intended to scare the victim, but knew there was a chance she would be killed); or "recklessly" (e.g., where the defendant should have known of the high risk that someone would be killed when he shot into the crowd). At trial, judges and jurors hear evidence of the defendant's behavior, which the prosecution hopes will establish both that the defendant had the opportunity to commit the crime and that he possessed the requisite mental states regarding the criminal harm caused to be found guilty. The prosecution may present evidence that the defendant was seen intentionally pointing the gun at the victim, or made prior threats against the victim; and evidence of motive (e.g., reasons the defendant would benefit from the victim's death) may be offered. Different levels of punishment may be warranted depending upon the level of mental state assigned: a purposeful homicide will result in more serious punishment than a reckless one.

Because the criminal law is couched in folk psychological terms, scientific evidence of a defendant's mental state, such as a medical diagnosis, evidence of a mental disorder, or fMRI or PET scan evidence, should only be presented to the court if such evidence undermines or supports attribution of a commonsense mental state at issue or if it supports an affirmative defense (e.g., mental incapacity or legal insanity), also couched in folk terms.[7] For example, evidence of a brain tumor is relevant to a defendant's guilt only if it indicates he did not possess the intentional mental states required under the law or to help establish that some excuse, such as diminished mental capacity, existed at the time of the crime.

For a defendant to be found guilty the prosecution must also prove a causal link between his culpable mental state and the criminal harm for

[6] (1985). Model Penal Code and Commentaries (Official Draft and Revised Comments). The Model Penal Codes was drafted to be used a model for state criminal codes, to increase the consistency of codes across the various U.S. states.

[7] One exception to this general rule may be the requirement of a "mental disease or defect" by the M'Naghten rule for legal insanity, which may require a mental disease be presented by an expert witness in terms of a scientific psychological theory.

which he has been arrested.[8] Certain very odd cases highlight the need to establish, and not assume, this link. Imagine a case where a defendant, Mike, intended to kill his neighbor, and Mike's behavior actually caused the neighbor's death. However, Mike had no intention of killing the neighbor at the time when he caused his neighbor's death: Mike set his own backyard shed alight, having no clue his neighbor was in there; indeed, Mike intended to kill his neighbor later that evening by poisoning him. In this case the actual death of Mike's neighbor was not causally related to Mike's intention to kill him, and Mike cannot be held responsible for murder (although he may have been negligent in the way he set fire to his shed).

The link between a defendant's mental state and the criminal harm must thus be thought of as a separate component of guilt. However, this link can be established in nonlinear ways, as happens with cases of "transferred intent." In a typical homicide case, the prosecution aims to prove that the defendant desired the victim's death and held the belief that some course of action would cause this death; and that the defendant then performed this action, which did indeed cause the victim to die. However, imagine a case where I intend to kill Anya, but my hand shakes so much that I kill Xander (who was standing next to her) instead. Under the doctrine of transferred intent, the court may transfer my purposeful intent to kill Anya to the criminal harm of Xander's death. In this atypical case, the criminal harm is causally linked to my culpable mental states because my desire to kill caused me to pull the trigger, but the law is willing to transfer the intent from the intended target for another. The situation becomes more difficult if Anya sees me pointing the gun at her and then pushes Xander into the path of the bullet, because my culpable intent to kill may no longer be considered the proximate cause of Xander's death. In this case Anya may also be considered culpable for Xander's death: if the court finds Anya acted knowing it was likely Xander would die, it seems clear her mental states are causally linked to Xander's death.

These examples are offered to emphasize how crucial mental state causation is to attributing criminal responsibility to defendants and thus, also to criminal punishment. It should be obvious that if it turned out to

[8] Criminal guilt typically requires that the defendant's behavior is the proximate cause of the criminal harm: this means that without the defendant's behavior, the criminal harm would not have occurred.

be the case that the mental states the criminal law seeks to attribute to defendants weren't *real* – if either (1) commonsense mental state concepts don't reliably refer to states or events within defendants or (2) such states or events fail to serve as a causal factor in criminal harm, then attributions of criminal responsibility by the criminal law are unjustified. As prominent legal scholar Stephen Morse notes, the law will be "fundamentally challenged" if it is shown that the commonsense psychology the law depends upon "is wrong and we are not the type of creatures for whom mental states are causally effective."[9] To put it bluntly: if mental causation isn't true, we are putting people in prison for no principled reason. Anyone hoping to vindicate the responsibility attributions made via the criminal law must seek a theory of the mind-body relationship that preserves the reality of commonsense mental states and their causal powers.

In the past 50 years or so there has been a lively philosophical debate regarding whether such a theory can be found. Critics of commonsense psychology, such as Paul and Patricia Churchland, expressed doubts about any theory of the mind–body relationship because they claim the mind (couched in folk terms) doesn't exist. Commonsense psychology, they argued, is unlikely to be true because the theory described entities in our heads (mental states), and these entities could not be directly perceived via the human senses – they were formed at a time when we at best had indirect access to the properties of brain entities or processes.[10] On the other hand, "realists" about commonsense mental states such as Jerry Fodor argued that the truth of commonsense psychology is evident from its predictive power. Even though the predictive power of commonsense psychology isn't completely accurate, it does seem indispensable to human existence, precisely because of how well it works.[11] Although commonsense explanations seem to work ceteris paribus, meaning that they aren't without exception, these ceteris paribus phrases are necessary to form mental state generalizations in all of the special

[9] Morse, 2011, 534.

[10] See Churchland, P. "Eliminativist Materialism and the Propositional Attitudes" (1981) *Journal of Philosophy*, 78; Churchland, P. *Neurophilosophy: Toward a Unified Science of the Mind-Brain* (Cambridge MA: MIT Press, 1986); Churchland, P. and P. S. Churchland *On the Contrary: Critical Essays, 1987–1997* (Cambridge, MA: MIT Press, 1999).

[11] Fodor, J. "Fodor's Guide to Mental Representation: The Intelligent Auntie's Vade-Mecum" (1985) 373 *Mind*, 76–100.

sciences, and we don't doubt the existence of the items studied by economics or political science.[12]

Recently the eliminative position has been reworked by some scholars to claim not that commonsense psychological concepts fail to refer, but that folk psychology assumes a particular type of mental causation that is false. For example, philosophers Derk Pereboom and Gregg Caruso argue that legal responsibility assumes a freedom of will or type of agential control that human beings do not actually have.[13] They conclude that because we do not have the required level of control over our actions, they cannot be considered free, and thus no one is ever responsible for their acts; similarly, no one ever really deserves criminal punishment.[14] Caruso concludes that dangerous actors should be "quarantined" instead.[15] Many compatibilists about free will disagree, of course, and some legal scholars have argued specifically that the criminal law is compatible with the truth of determinism and its descendants.[16]

Even for those who believe that commonsense mental state concepts do truly refer, and that folk psychology describes a causal process that can ground responsibility, the task remains to posit a relationship between mental states couched in commonsense terms and their physical nature such that mental causation is possible. The nature of this relationship depends crucially on a theorist's ontological view of world.

[12] Fodor, J. "Special Sciences (or the Disunity of Science as a Working Hypothesis)" (1974) 28 *Synthese*, 97–115.

[13] See, Caruso, G. "Free Will Skepticism and Criminal Behavior: A Public Health-Quarantine Model" (2016) 32 *Southwest Philosophy Review* and Pereboom, D. "Free Will Skepticism and Criminal Punishment" in T. Nadelhoffer ed., *The Future of Punishment* (New York: Oxford University Press, 2013), 49–78.

[14] See Pereboom, D. and G. Caruso "Hard-Incompatibilist Existentialism: Neuroscience, Punishment, and Meaning in Life" in G. Caruso and O. Flanagan (eds.), *Neuroexistentialism: Meaning, Morals, and Purpose in the Age of Neuroscience* (Oxford: Oxford University Press, 2018).

[15] Caruso, 2016.

[16] See Morse, S. "Lost in Translation?: An Essay on Law and Neuroscience" (2011) *Law and Neuroscience, Current Legal Issues*, 13 and Morse, S. "Neuroscience, Free Will, and Responsibility" in W. Glannon ed., *Free Will and the Brain: Neuroscientific, Philosophical, and Legal Perspectives* (Cambridge: Cambridge University Press, 2015). Many now think universal determinism is likely to be false because quantum mechanics indicates that much of what happens in the universe is indeterminate (Pereboom and Caruso, 2018 at 9). But of course the notion that our actions may result from indeterminate events makes agential control even less likely than the notion that our actions are related to determined events.

For scholars committed to physicalism – the idea that only one thing exists in the world, and it is physical stuff – the question posed is how to understand the way in which different epistemic entities (entities described under a scientific theory, and entities described under the folk psychological theory) are related.[17] If everything that exists is physical stuff, mental states are physical stuff. The question is what sort of stuff they are, and how we are to understand the relationship between the laws that govern mental states under a physical description, versus the rational relationships they seem to exhibit as folk psychological explanations.

To sum up this section: criminal verdicts crucially depend upon *mental causation*, the notion that our mental states can cause things to happen in the world. I will argue later that only certain theories regarding the relationship between mental states and physical states (such as brain states) preserve mental causation. Our best candidate theory for preserving mental causation, and the practice of attributing criminal responsibility to defendants, and punishing them, is a non-eliminative reductive theory of mental states. But first, I will argue in the next section that the two forms of eliminativism mentioned earlier are false.

I will only discuss theories that claim to be compatible with physicalism. This is because I think the problems ontological dualism poses for mental causation are well known and have been dealt with thoroughly elsewhere. I will not attempt to offer a complete account of any theory discussed; but instead hope to highlight the difficulties each theory poses for mental causation, with the idea that I might show my favored theory, non-eliminative reductionism, is the best theory to preserve mental causation, and thus the best theory to ground responsibility attributions made in the criminal law.

[17] As Ney notes: "Physicalism is, if not a received view, at least a starting point in much of philosophy today. There are those who dissent, but the view that physics, perhaps even a physics of the not-too-distant future, may provide an exhaustive guide to the fundamental constituents of our world, a view rejecting fundamental mental and biological phenomena, is widely adopted as a place to begin debates about fundamental ontology, the nature of mind, the special sciences, and even the status of ethics and normativity." Ney, A. "Microphysical Causation and the Case for Physicalism" (2016) 57 *Analytic Philosophy*, 141–164. For an argument in support of physicalism, see Papineau, D. *Philosophical Naturalism* (Oxford: Blackwell, 1993).

Elminativism

Eliminativism about Commonsense Psychological Concepts

If eliminativism about mental states is true, criminal verdicts generated by looking for particular mental states in defendants, causally related to criminal harm, would be no different than criminal verdicts generated via phrenology or by throwing the defendant into a vat of water to see if she sinks or swims. This brand of eliminativism could not possibly preserve mental causation because it does not preserve commonsense mental concepts. If mental state concepts must be eliminated, attribution of criminal responsibility and punishment ought to be suspended.

Paul and Patricia Churchland, the most well-known defenders of eliminativism in the 1980s and 1990s, offer several arguments that commonsense psychology fails to refer to real entities in the world and must be eliminated. The Churchlands have claimed commonsense psychology (1) has not "shown the expansion and developmental fertility one expects from a true theory"; (2) shows no signs of being smoothly integrable with "... the emerging synthesis of the several physical, chemical, biological, physiological, and neurocomputational sciences; and (3) fails to "... to explain a considerable variety of central psychological phenomena" like mental illness, memory, intelligence differences, and the different forms of learning. A viable theory, they argued, would show more progress, be more easily integrated with scientific theories, and would not have such explanatory gaps.[18]

However, close examination of these arguments reveals they do not provide support for the idea that commonsense psychology posits concepts that fail to refer to anything in the world. Indeed, there seems to be good evidence that commonsense psychological concepts do reliably refer to states or events in our heads: as indicated earlier, such concepts actually work very well to explain our behavior.[19] If mental state concepts did not refer to fairly stable entities in our heads, that in turn had fairly reliable causal effects, how could such concepts be used so successfully to predict human behavior? For example, if thirst did not exist and thus failed to have a stable causal identity, how might we accurately predict – all things being equal – that persons who are thirsty will go get a drink?

[18] See Churchland and Churchland, 1999.

[19] Fodor, J. *Psychosemantics: The Problem of Meaning in the Philosophy of Mind* (Cambridge, MA: MIT Press, 1987).

In sum, as Fodor has noted, one big reason to think commonsense psychology isn't an utterly false theory is that it *works so well*.[20]

Another way to judge the veracity of commonsense psychological concepts is to look to see whether their referents are indeed emerging from progress made in scientific psychology. Does our best current scientific psychology appear to vindicate or debunk the notion that perceptual states like pain and thirst, emotional states like anger, and informational states like the belief that it is raining, have existence as brain states? In the very least, evidence generated by cognitive science and neuroscience indicate that mental states such as pain, anger, thirst, and beliefs do indeed exist and are instantiated in the brain.[21] The search for the functional identity of such commonsense mental states seems to be well underway; some might even agree that it is progressing nicely.

Given this, it isn't clear why the Churchlands seemed convinced that commonsense psychology shows no signs of being integrable with the hard sciences. Insofar as there is progress in neuroscience, this progress is made via discovering the referents of commonsense mental concepts. This is because the primary way to search for functional capabilities – or dysfunction – in the brain is by attempting to locate the referents of commonsense concepts in the brain. Neuroscientists use commonsense concepts to describe a cognitive capacity such as facial recognition (a capacity that can be understood as the ability to form beliefs that something is a face) before they go looking for that capacity in the brain. They then seek the capacity or incapacity by giving subjects cognitive tasks such as identifying faces – and indicating they see a face using commonsense terms – while, for example, the subjects are in an fMRI machine ("Yes, I see a face now," and "Nope, I don't see a face.").

Further, regarding the Churchlands' claim that an "emerging synthesis" regarding the nature of the brain is currently under construction by

[20] Ibid. at 3–4.

[21] To offer just a few examples: Egan, G. et al. "Neural Correlates of the Consciousness of the Emergence of Thirst" (2003) 100 *Proceedings of the National Academy of Sciences of the United States of America*, 15241–15246; Denson, E. F. et al. "The Angry Brain: Neural Correlates of Anger, Angry Rumination, and Aggressive Personality" (2009) 21 *Journal of Cognitive Neuroscience*, 734–744. It is unclear what sort of information discovering neural correlates really gives us about the nature and content of commonsense mental states. But most would agree that this sort of evidence indicates it is likely such mental states have some sort of physical instantiation in the brain.

the hard sciences: (1) this may be true, but a close examination of the progress made in brain science also shows slow progress toward an understanding of human cognitive capacities couched in commonsense psychological terms and (2) such a synthesis itself would not prove commonsense psychological concepts to be false. Instead, it would seem to show that commonsense concepts are real in that they reliably refer to the brain states we already know are causally linked to human behavior. To show commonsense concepts are radically false, eliminativists would be required to show that such concepts are in fundamental conflict with the way the hard sciences describe the world. But again, progress in brain science does not indicate this. Instead, progress is being made to identify the referents of mental states in the brain.

This is not to say that we should expect commonsense psychology to be a perfect, or even very accurate descriptor of the brain states that cause human behavior. As I will discuss in more detail later, commonsense concepts are instead likely to roughly refer to brain processes. And it ought not surprise us that commonsense psychology fails to adequately explain certain psychological phenomena, such as the way in which memories are made and stored. The apparent function of mind-reading and commonsense psychology is (at least in part) to understand and predict human behavior, which would require identification of mental states as brain states that have enough of a discrete identity to have reliable causal effects. This, in turn, would require recognition that persons can and do indeed form memories. However, it may not be necessary for the folk to understand how the cognitive processes vital to memory creation and storage operate. But it hardly follows from the fact that a theory is incomplete that the theoretical claims that are made are false.

To put it plainly, it seems the Churchlands' call to eliminate commonsense psychology as a theory of human behavior was imprudent. And this ought to be a relief to legal scholars, responsibility theorists, and anyone else who supports the project of holding persons responsible. This is because, as stated earlier, if commonsense mental states aren't real – in the same way that witches aren't real – then we are unjustified in finding persons guilty of a crime. The courts, in assessing criminal guilt, would be seeking fictional/nonexistent entities that (as such) cannot possibly have any impact on human behavior. If commonsense psychological concepts are eliminated, then the mental states the court is looking for in order to attribute criminal responsibility do not exist.

Elminativism about Free Will

Most philosophers now agree that Churchland-style eliminativism was ill-conceived. However, a subtler brand of eliminativism relevant to criminal responsibility has recently emerged: eliminativism about free will.[22] These eliminativists hold that free will doesn't exist, and free will is required for criminal responsibility; therefore, no one is ever criminally responsible. The position is different from that of the Churchlands, in that the new eliminativists do not advocate for elimination of common-sense mental states (in favor of some other, scientific understanding of the causes of human behavior). Instead they argue that our common-sense or folk responsibility practices, which underpin the criminal law, assume a special sort of causal power on the part of agents – a sort of agential control over their decisions and actions – that they do not actually have. In this case, argue the new eliminativists, it isn't fair or just to hold persons responsible for harm caused by such actions, or subject them to harmful punishment, because they aren't really the cause of such harm.

This is an interesting argument from the perspective of the mind body relationship, for it seems that the new eliminativists hold that mental states are indeed real things in the world.[23] However, our mental states do not have causal powers such that we can blame and hold people responsible for their actions. The folk notion of mental causation is radically false and ought to be eliminated as an explanation of human behavior. Once this is done, responsibility, including criminal responsibility, disappears.

Some have argued that the new eliminativist position is supported by experimental research into ordinary perceptions of free will. Certain studies seem to indicate that the ordinary folk don't tend to believe determinism is true, and when they are forced to consider the possibility that the universe is deterministic, the folk are less likely to apply praise and blame.[24] When presented with an explicit account of a determinate

[22] See Pereboom, n.13; Caruso, n.13; Caruso, G. *Free Will and Consciousness: A Determinist Account of the Illusion of Free Will* (Lanham, MD: Lexington Books, 2012); Caruso, G. "Free Will Eliminativism: Reference, Error, and Phenomenology" (2015) 172 *Philosophical Studies* 10, 2823–2833; Pereboom, D. *Living without Free Will* (New York: Cambridge University Press, 2001).

[23] See Caruso, ibid, 2012 and Pereboom, n.13.

[24] Nichols, S. and J. Knobe "Moral Responsibility and Determinism: The Cognitive Science of Folk Intuitions." (2007) 41 *Nous* (4), 663–685.

universe, one study showed that persons were hesitant to hold persons within that universe responsible.[25] However, other studies seem to have found that persons are still willing to hold agents responsible, even if those agents' actions appear to be determined.[26] Murray and Nahmias attempt to explain the conflicting data as confusion on the part of the folk regarding what determinism really means. This confusion, they say, fueled incompatibilist intuitions in cases where the study subjects were hesitant to attribute responsibility. If subjects are led to believe that determinism means that a person's conscious mental states are bypassed as a cause of their behavior, then responsibility intuitions are undermined. However, if subjects feel that an action is related to or caused by a person's conscious mental states, subjects seem willing to attribute responsibility, even if determinism is true. Because determinism does not entail bypassing the conscious self as a causal agent, Murray and Nahmias claim folk intuitions with regard to responsibility may remain intact even if determinism is true.[27]

It seems likely that folk intuitions about free will are mixed: that they are partly incompatibilist (hold determinism is incompatible with free will), and partly compatibilist (hold determinism is compatible with free will). Thus, if determinism turns out to be true, folk notions of free will may be partly true and partly false.[28] Nahmias and Murray's arguments, however, indicate that the folk notion of free will is very likely to be true enough to be preserved, despite determinism. A fairly liberal theory of reference would certainly seem to allow revision of the folk concept of free will such that it could be construed to refer to real brain processes and events corresponding to the conscious mental states that play a causal role in behavior. In this way the folk concept of free will might be more similar to the concept of *whale* – which had to be revised because humans initially assumed they were fish – than the concept

[25] Nichols, S. and A. Roskies "Bringing Moral Responsibility Down to Earth" (2008) 105 *Journal of Philosophy* (7), 371–388.

[26] Nahmias, E. and D. Murray "Experimental Philosophy on Free Will: An Error Theory for Incompatibilist Intuitions" in J. Aguilar, A. Buckareff, and K. Frankish eds., *New Waves in Philosophy of Action* (London: Palgrave-Macmillan, 2010) 189–215 and Murray, D. and E. Nahmias "Explaining Away Incompatibilist Intuitions." (2014) 88 *Philosophy and Phenomenological Research* (2), 434–467.

[27] Murray and Nahmias, ibid.

[28] See Vargas, M. *Building Better Beings: A Theory of Moral Responsibility*, (Oxford: Oxford University Press, 2013).

witch, which had to essentially be eliminated because it turned out there were no women with magic powers.

Gregg Caruso and Derk Pereboom, however, have argued the opposite, claiming that the folk concept of free will is more like the concept *witch* than the concept *whale*. Two arguments regarding the radical falsity of folk notions of free will seem to be present in their work. First, Caruso claims free will historically refers to the *phenomenological feeling* of free will, which is essentially libertarian and incompatibilist in nature.[29] That is, the feeling of making a free decision indicates that our decisions are not determined (or subject to indeterminate forces). I disagree, however, that this feeling of freedom is central to the folk notion of free will, and I think my view is supported by the relatively small role the phenomenal feeling of freedom plays in responsibility assessments, a human practice that robustly engages folk notions of free will. The criminal law, for example, does seem to engage folk notions of free action, but in what sense? A defendant is not criminally responsible if the act was subject to duress, hypnosis, very low IQ, sleepwalking, or acts committed under the influence of a serious mental illness, and these are all circumstances where we might consider an agent's freedom to act impaired. But in each of these cases the defendant's cognitive system is either deviant when compared to a normal adult, or the defendant is constrained by very abnormal external circumstances. Only in the first case of duress does excuse correlate with a lack of feeling free. (I assume a hypnotized person and a sleeping person either feel free in their actions or feel nothing – I do not think they feel unfree. And I am quite sure some persons suffering from serious mental illness feel free.) Others who we excuse from full criminal culpability, including young children and those with low IQs, also feel quite free as they act, but their cognitive incapacities relative to a normal adult result in lesser culpability under the law. Further, many of those determined to be fully criminally responsible sometimes do not feel free: we often hold persons responsible for acts committed under serious emotional stress or under the pressure of a very weighty decision or very difficult circumstances, even though in such cases the agent may feel unfree.

In a later coauthored paper, Pereboom and Caruso offer a second argument against the folk concept of free will by claiming it refers to a type of agential control that persons do not have regardless of whether

[29] Caruso, 2015.

the universe turns out to be largely deterministic or somewhat indeterministic. They argue both compatibilism and libertarianism about free will are false because neither provides an account of agential control sufficient to vindicate the truth of free will. "Against the view that free will is compatible with the causal determination of our actions by natural factors beyond our control, we argue that there is no relevant difference between this prospect and our actions being causally determined by manipulators. Against event causal libertarianism, we advance the disappearing agent objection, according to which agents are left unable to settle whether a decision occurs and hence cannot have the control required for moral responsibility."[30]

I think Pereboom and Caruso are right to claim the folk notion of free will refers at least in part to agential control over action. Again, however, it seems that the folk concept of free action may refer to the causal efficacy of conscious mental states as the locus of this control. Agents feel very strongly aligned with their conscious values, plans, and decisions, and such values, plans, and decisions are indeed causally related to much of human behavior.[31] And, the folk notion of freedom present in the criminal law seems clearly aimed at these types of causally efficacious mental states (classified as "purposely" and "knowingly," etc.). As mentioned earlier, cases of legal excuse tend to be cases where there is something wrong with the way in which these conscious states are formed and executed. For example, in cases of mental illness, addiction, and sleepwalking the states causing action are "abnormal" because they are not subject to the typical conscious cognitive review process, and in the case of young children and those with very low IQs, this system of review is underdeveloped.[32]

[30] See Pereboom, 2001; Caruso, 2012; and Pereboom and Caruso, 2018.

[31] This is not to deny that there are many unconscious mental factors that influence behavior. However, there are theories – my colleagues and I have a scientifically based version, and Angela Smith has a purely philosophical one – that the possibility of conscious rational endorsement is enough to assign responsibility to the mental cause of an act. See Smith, A. "Attitudes, Tracing, and Control" (2015) 32 *Journal of Applied Philosophy* (2), 115–132.

[32] See Fagan, T., W. Hirstein, and K. Sifferd "Innocent Minds: Child Soldiers, Executive Functions, and Culpability" (2016) 16 *International Criminal Law Review* (2), 258–286; Sifferd, K. (2012). "Translating Scientific Evidence into the Language of the Folk: Executive Function as Capacity-Responsibility" in N. Vincent ed., *Legal Responsibility and Neuroscience* (Oxford: Oxford University Press, 2012); Sifferd, K., W. Hirstein, and T. Fagan "Legal Insanity and Executive Function" in M. White ed., *The Insanity Defense: Multidisciplinary Views on Its History, Trends, and Controversies* (Santa Barbara, CA:

Caruso and Pereboom say that conscious causing isn't sufficient to ground agential control because it is no different than being causally influenced by intentional manipulators. It is well established that our conscious mental states are influenced by a whole host of factors outside the agent's control,[33] and this includes the possibility of malevolent manipulators. But it isn't clear that the fact that agents are influenced by outside forces is relevant to our assessment of the reference of the folk notion of free action. Caruso and Pereboom may feel the folk fail to distinguish between "normal" outside influences, like parenting and schooling, and "abnormal" influences, such as the manipulations of evil neuroscientists. But the folk concept of free action does not need to operate in a manner considered fair or consistent by philosophers for it to truly refer. The folk notion of "tasty" is hardly consistent (in the sense of treating like cases alike) with result to outcome: my concept of "tasty" successfully refers to strawberries, pixie sticks, and tequila, and one can hardly identify anything these tasty items have in common except the fact that I find them tasty. But this doesn't mean the concept of "tasty" ought to be eliminated: it truly and reliably refers to things in the world and refers accurately enough for us to use the concept to understand and predict our own and others' behavior. Philosophical arguments that a folk concept like "tasty" or "free action" fails to treat like cases alike may indicate a need to revise our understanding of the referents *in order to achieve certain social or ethical goals*. However, such an argument cannot provide grounds to eliminate either concept.

I believe that our responsibility practices are fair in large part, and that folk concepts are well equipped to treat like cases alike with regard to human action, but I need not argue for this here. All I need to argue to defend against an eliminative challenge is that the folk concept of free action does indeed reliably refer to a category of human behavior, and there is good proof of this in the criminal law. The folk concept of free action appears to refer – at least in part – to action causally related to the conscious mental states of a normal adult's cognitive system.

Praeger, 2017), 215–242; Hirstein, W. and K. Sifferd. "The Legal Self: Executive Processes and Legal Theory" (2011) 20 *Consciousness and Cognition* (1), 156–171.

33 See, for example, the enormous amount of research that has been done on the way in which "situational factors" impact persons' decisions.

Many legal scholars have similarly argued that responsibility practices of the criminal law do not demand libertarian or other false notions of free will or agential control.[34] As Stephen Morse claims:

> [T]he folk-psychological model of the person that is central to our explanations of human behavior and to responsibility doctrines and practices is not challenged by determinism in general or by neurodeterminism in particular. Criminal responsibility doctrines and practices are fully compatible with the truth of determinism (or causal closure). Until science conclusively demonstrates that human beings are not responsive to and cannot be guided by reasons and that mental states do not play even a partial causal role in explaining behavior, the folk-psychological model of responsibility will endure as fully justified.[35]

Morse agrees that legal responsibility doesn't rest upon whether a defendant felt free, or on a special kind of agential control. Instead, the law holds persons responsible who cause harm via their own conscious mental states. Folk concepts underpinning criminal responsibility also seem to track what Morse calls "minimal rationality"[36] and what philosophers call "reasons-responsiveness."[37] Some have argued that consciousness is what enables this rational capacity in human beings. When a mental state becomes conscious, it can be subject to review by the agent's long-term values and desire.[38] Morse argues that the criminal law

[34] See: Morse, S. "Criminal Responsibility and the Disappearing Person" (2007) 28 *Cordozo Law Review*, 25–45; Morse, S. "Determinism and the Death of Folk Psychology: Two Challenges to Responsibility from Neuroscience" (2008) 9 *Minnesota Journal of Law, Science and Technology* (1), 1–36; and Morse, S. "Neuroscience, Free Will, and Responsibility" in W. Glannon ed., *Free Will and the Brain: Neuroscientific, Philosophical, and Legal Perspectives* (Cambridge: Cambridge University Press, 2015).

[35] See Morse, 2015 at 253.

[36] Morse says, "The legal view of the person does not hold that people must always reason or consistently behave rationally according to some preordained, normative notion of rationality. Rather, the law's view is that people are capable of acting for reasons and are capable of minimal rationality according to predominantly conventional, socially constructed standards. The type of rationality the law requires is the ordinary person's common-sense view of rationality, not the technical notion that might be acceptable within the disciplines of economics, philosophy, psychology, computer science, and the like." Morse, ibid. at 256.

[37] Fischer, J. M. and M. Ravizza *Responsibility and Control: A Theory of Moral Responsibility* (Cambridge: Cambridge University Press, 1999).

[38] Neil Levy argues that for an agent to be held responsible for an action it must be causally related to conscious mental states; Levy, N. *Consciousness and Moral Responsbility* (Oxford: Oxford University Press, 2014). However, my colleagues and I argue that culpable action must be causally related to states that are *available* to conscious executive

seems to reflect the standard compatibilist position: that some subset of determined actions/actions influenced by factors outside the agent can be assigned to the agent such that she may be held responsible for them. At least from the perspective of legal responsibility practices, the free will that matters is the capacity to act in accordance with reasons, including moral and legal reasons.

It seems, at least as an argument for eliminativism about mental causation, the Caruso and Pereboom position fails if the folk consider free action that action caused by conscious mental states (or mental states available to consciousness) such that such states are related to a minimally rational, or reasons-responsive, cognitive system.[39] Even if some of the folk intuitions refer to libertarian free will and thus are incorrect, it seems the core aspects of the concept are probably correct – especially, the aspects that seem especially important to the criminal law – and thus the concept of free will need not be eliminated. A whale may not be a fish, but it is still a very large gray animal that lives in the ocean.

Both forms of eliminativism discussed earlier – about mental states and agential causation – offer a theory of the mind–body relationship that poses a threat to mental causation. If mental states aren't real, they can't cause behavior. Further, if agents don't cause behavior in the way the folk think they do, then mental causation is false, and persons cannot be held responsible for their acts. To put it another way, if, after exploring the eliminativist arguments, we were convinced that either version of eliminativism was true, it would be hopeless to seek a theory of the mind–body relationship that preserves mental causation. However, it seems that both eliminative threats can be avoided: neither the folk concepts of mental states, nor the folk concept of free will, are radically false. Thus, such concepts can be preserved to ground responsibility attributions made in the criminal law. Our task now is to seek a positive theory of the mind–body relationship that preserves the mental as a real thing in the world with causal powers. This theory should help us understand how the mental possesses causal powers via its physical realizers.

review – they needn't necessarily be conscious themselves. See Hirstein, W. and K. Sifferd, n. 32, 2011; Sifferd, n. 32, 2012; and Fagan, Hirstein, and Sifferd, n. 32, 2016.

39 See Levy, ibid.

Non-Reductive Physicalism/Emergence

Before discussing non-reductive physicalism, let me say a few words about why I have dismissed ontological dualism as a possible theory of the mind–body relationship that can preserve mental causation. Ontological dualism – the idea that the mental is of fundamentally different stuff than the physical – poses well-documented problems for mental causation.[40] If mental states occupy another ontological category than the physical body, it becomes very difficult to explain how the immaterial mind has physical causal effects.[41] As Princess Elizabeth of Bohemia commented to Rene Descartes in a letter in 1643, there seems to be no way to conceive of how an immaterial soul or immaterial mental states can have any effect on material substances, including the material substance of the human body. "For the determination of movement seems always to come about from the moving body's being propelled – to depend on the kind of impulse it gets from what sets it in motion, or again, on the nature and shape of this latter thing's surface. Now the first two conditions involve contact, and the third involves that the impelling thing has extension; but you utterly exclude extension from your notion of soul, and contact seems to me incompatible with a thing's being immaterial."[42]

Philosopher Colin McGinn has said the following about dualism about the mind:

[40] See Davidson, D. "Actions, Reasons, Causes" (1963) 60 *Journal of Philosophy* 685–700. See also Mele, A. *Springs of Action* (New York: Oxford University Press, 1992).

[41] See Kim, J. "The Myth of Nonreductive Materialism" in P. K. Moser and J. D. Trout eds., *Contemporary Materialism* (New York: Routledge, 1995), 134–149; Sifferd, K. "What Does It Mean to Be a Mechanism? Stephen Morse, Non-Reductivism, and Mental Causation" (2014) *Criminal Law and Philosophy*, 1–17. Theories that deny an ontological reduction are thought by many to render the mental epiphenomenal. Due to the causal closure of the physical realm, physical states seem to be the obvious winners if the mental is forced to compete with regard to the causation of behavior. See Papineau, D. *Thinking about Consciousness* (Oxford: Oxford University Press, 2002). If commonsense mental entities or events cannot be understood or explained in terms of their physical underliers, then we cannot reliably attribute causal powers to them. The claimed epistemological gap threatens assignment of causal powers to the mental because such an attribution of causal powers seems excluded by the assignment of causal powers to the physical underliers. See Kim, J. *Physicalism, or Something Near Enough* (Princeton, NJ: Princeton University Press, 2005).

[42] Quoted in: Anscombe, E. and P. Geach *Descartes: Philosophical Writings* (Indianapolis, IN: Bobbs-Merrill Company, 1954).

> Dualism proposes to give the mind its ontological due, but the problem is that it has difficulties organizing a rendezvous between the two spheres: how does the mind affect the brain and the brain the mind? Whence the systematic correlation and interaction? . . . Dualism makes the mind too separate, thereby precluding intelligible interaction and dependence.[43]

As McGinn indicates, no sound theory has emerged explaining how an immaterial substance could create physical causal effects, and thus provide a plausible account of mental causation. However, McGinn thinks dualism gives the mind "its ontological due" because it holds what McGinn, and some other philosophers and legal theorists, feel must be true about the mind – that it cannot *just be* physical states or processes; that instead, there must be some important, special aspect of the conscious human mind that is separate from, and thus cannot be reduced to, its underlying physical realizers.[44] Even among philosophers committed to physicalism, this notion of the fundamental separateness of the mental from the physical seems salient, and has fueled non-reductive positions about the mind that have gained popularity in the past few decades. To my mind, the appeal of such theories is that they appear to allow theorists to have their cake and to eat it too: the mind remains "special" such that it cannot be reduced to the physical realm, but it doesn't occupy a new (nonphysical) ontological category. Non-reductive physicalism thus seems to represent a sort of metaphysical middle position.

The question posed here is what sort of theory of the mind–body relationship will preserve the mental causation required for criminal verdicts. My tactic in this section will be to closely examine two non-reductive theories that posit a *close* relationship between the mental and the physical: Donald Davidson's non-reductive physicalism and John Searle's biological naturalism. Searle argues that conscious states cannot be ontologically reduced to physical states, and Davidson argues that mental states resist an epistemological reduction. If these two non-reductive theories fail to preserve mental causation, despite the close relationship they posit between the mental and physical,

[43] McGinn, C. "All Machine and No Ghost?" *New Statesman*, 2012.

[44] This sentiment seems to fuel Stephen Morse's acceptance of non-reductive physicalism. He holds that mental states are not reducible because they "are realized in the brain – the mind-brain – but not at the level of neurons," Morse, n. 3, 2008: at 33.

I don't think we should hold out much hope that other non-reductive theories will fare any better.

Non-reductive physicalism is the position that although everything that exists is physical, and thus the mental is physical, some aspect of the mental cannot be reduced to the physical. Thus, to explore non-reductive theories about the mental requires exploration of the notion of reduction. Generally, reductions may be ontological or epistemological. Ontological reductions can be made between entities, objects, events, processes, and properties, and epistemological reductions are made between concepts, theories, models, and frameworks.[45] An ontological reduction is made between real-world entities, and a strong claim about ontological reduction would indicate that everything in the world can be said to be nothing more than or determined by some fundamental constituent of reality (e.g., atoms or the entities posited by quantum physics). An epistemological, or theoretical reduction might entail fully explaining or understanding higher-order mental states or events in terms of a science of lower-level brain states or events.

A non-reducible entity, property, theory, or framework is often said to be *emergent*, which is just another way of saying that it cannot be reduced. Thus both dualist and non-reductive positions on the mind–body relationship posit that some aspect of the mental is emergent and cannot be reduced to the brain. Davidson's brand of non-reductive physicalism is best understood as acceptance of ontological reduction and denial of an epistemological/representational reduction.[46] Searle's non-reductionism, on the other hand, denies an ontological reduction, while at the same time, trying to maintain its status as a physicalist theory. As we shall see, this isn't easy. One might wonder: if some aspect of mind occupies a separate ontological category from that of the physical brain or body, what sort of (physical) thing could it be?

Davidson's Non-Reductionism

Donald Davidson's non-reductionism is perhaps the most influential version of the theory. Davidson began to argue for a non-reductive physicalist position in the late 1960s and 1970s, partly in reaction to

[45] See Van Gulick, R. "Reduction, Emergence and Other Recent Options on the Mind/Body Problem: A Philosophic Overview" (2001) *Journal of Consciousness Studies* 8 (9–10), 1–34 at 3.4.

[46] Van Gulick, ibid.

the perceived failures of the type-identity reductive project, which aimed to solve the problem of mental causation by identifying types of mental entities with physical types, such as brain states.[47] Davidson argued the relationship between the mental and the physical is not that of identity (where a mental state *just is* some brain state or other) but instead argued for a relationship of weak supervenience. Specifically, Davidson argued that the mental supervenes on the physical, such that any change in a mental property has to be accompanied by some change in a physical property. However, the reverse does not hold: a change in physical properties does not necessarily entail a mental change.[48] This means that although each mental event has some physical identity or other, the same mental event could be instantiated as very different physical brain events: a mental state such as thirst is "token" identical to some physical state, but it may be identical with different physical states within and across persons.

The motivation for Davidson's position is "anomalous monism." Although all events occur within the single physical universe (the position of monism), physical and mental events seem to participate in very different types of laws. While scientific theories or frameworks seemed to posit entities governed by strict causal physical laws, the folk concepts and theories utilized when positing mental events do not seem so governed.[49] For example, I don't always go get a drink when I am thirsty – I might not get a drink if I were in the middle of teaching a class, or if I were running a race. Therefore, the mental and physical realms were deemed by Davidson to be incompatible insofar as there could not be strict psychophysical laws: explanations utilizing the ceteris paribus laws governing mental events could not be understood in terms of physical events, which appeal to the laws of physics. In other words, the laws of the special sciences (sciences focused on understanding human behavior) are so fundamentally different from those governing other things in the world that mental explanations of behavior cannot be reduced to (read: understood in terms of) physical explanations of behavior.[50]

Davidson's version of non-reductionism has been criticized on various grounds, but the theory is particularly worrying with regard to mental causation. The problem with Davidson's theory is that if the same mental state can supervene on different physical states, the mental

[47] See Smart, J. "Sensations and Brain Processes" (1959) *Philosophical Review* 66, 141–156.
[48] See Davidson, D. *Essays on Actions and Events* (Oxford: Clarendon Press, 1980).
[49] Davidson, ibid. [50] Davidson, ibid.

cannot reliably substantiate causal properties because these properties are instantiated within the physical realm. As Jaegwon Kim has noted, Davidson's position means that the very same network of physical causal relations would obtain even if one were to redistribute mental properties over it randomly; on Davidson's theory "... you would not disturb a single causal relation if you ... removed mentality entirely from the world."[51]

Davidson acknowledges that the physical realm operates in accordance with strict causal laws, and that physical states are causally related to behavior. So, when I get up to get a drink, there is a physical causal explanation for my action that adheres to strict laws. There is, of course, also a mental explanation: I was thirsty; I knew there was a drinking fountain in the hallway; thus, I decided to go get a drink of water. But at the heart of Davidson's theory is the claim that the mental state of being thirsty can be instantiated in different ways in the brain from one tokening to the next. So my thirst and belief that there is a drinking fountain in the hallway might be instantiated in my, say, in my prefrontal cortex in one moment, but the next time I am thirsty, this thirst supervenes on my amygdala; and and my belief that there is a drinking fountain in the hallway supervenes on my visual cortex.

However, if this were the case (1) it isn't at all clear how the mental state of thirst can have reliable behavioral effects, such that if you know I am thirsty you might predict I was going to get a drink, and (2) it isn't clear how the mental explanation is a causal explanation of my behavior, given that the physical explanation is most certainly causal, and the mental explanation only attaches to the physical explanation in an unprincipled way (or, to put it plainly, willy-nilly). In sum: the physical causal nature of the brain is indeed playing a primary causal role in our behavior, and if we cannot reliably link mental states to these causal processes, the mental does not act as a cause of our behavior. Thus, on Davidson's theory the mental ends up looking as though it is epiphenomenal (has no causal power). But from the perspective of the law, when judges or juries look for the mental states that caused criminal harm, they are looking for a causal relationship between mental event and criminal harm that doesn't reliably exist. For this reason, Davidson's theory does not posit a relationship between the mind/brain that can support criminal verdicts.

[51] See Kim, n. 41, 1995 at 136.

Searle's Non-Reductionism

I have discussed Searle's non-reductionism in detail elsewhere.[52] Here I will attempt to give a short summary of some of my prior arguments. Searle's non-reductionism is interesting with regard this chapter's project because one of Searle's expressed aims in articulating this theory is to preserve mental causation. Searle's theory, termed "biological naturalism," consists of four theses:[53]

1. Consciousness cannot be reduced to its neurobiological basis because "such a third-person reduction would leave out the first-person ontology of consciousness."
2. Conscious states, however, are causally reducible to neurobiological processes – they are not something "above and beyond" such processes.
3. Portions of the brain system composed of neurons are conscious, although individual neurons are not.
4. Conscious states function causally – they have physical effects.

The causal efficacy of mental states (thesis four) relies upon their causal reducibility (thesis two). Thus, Searle says conscious mental states have physical effects by virtue of the place they occupy within a physical causal system: the causal properties of a conscious state are identical with the causal properties of the physical state that realizes it.[54]

Searle maintains a non-reductive position despite his claim that mental entities are causally reducible by arguing that the *conscious phenomenal* properties of a mental state are emergent.[55] The very point of the concept of consciousness is to capture the first-person, subjective experience, claims Searle, and this subjective phenomenal feel of a conscious state is not constituted by, or identical to, underlying brain states or processes. Instead, phenomenal feel is an ontologically emergent property of biochemical brains like ours. Thus, while Davidson admits an ontological reduction and denies an epistemological one, Searle denies an ontological one, at least with regard to phenomenal feel.[56]

[52] See Sifferd, 2014, n. 41.
[53] Searle J. *Mind: A Brief Introduction* (New York: Oxford University Press, 2004).
[54] Ibid. at 118. [55] Ibid.
[56] Ibid. Again, an ontological reduction involves a claim that an object or event of a certain type can be shown to be nothing more than or determined by objects or events of another type, in the way that a rock can be shown to be nothing more than atoms arranged in a certain way.

Searle claims that consciousness is a "causally emergent property" of systems of neurons, whose existence can be explained by the causal interactions between elements of the brain at the microlevel, but can only be accessed by the person who owns the brain producing consciousness.[57] But this argument seems odd given that Searle denies property dualism – that is, he denies that there are nonphysical properties. This means that the properties of consciousness must be physical spatial properties of the world with causal effects. But in this case, it seems exceedingly mysterious how certain physical and spatial properties of the brain that have physical causal effects can be accessible only to consciousness itself.[58] In other words, Searle's position is ontologically very odd: it is unclear how such phenomenal feels, which he admits are physical causal properties, are properties that are necessarily invisible to science.

Of course, many would agree with Searle that empirical examination of a brain state (using current scientific means) will leave out information of how this brain state feels to the holder of the state. But this is an argument against epistemological, not ontological, reduction. And as Van Gulick notes, one oughtn't to infer that mental properties are ontologically emergent even if it turns out to be the case that we cannot representationally reduce our mental concepts or theories to physical ones.[59] To do so would be to assume an epistemological divide is enough to prove an ontological one. And this is what Searle seems to do: he resists an ontological reduction based on his notion of a subjective/objective divide. This move seems especially suspect given that Searle causally identifies mental states with physical states. One important way to divvy up ontological entities is via their causal powers. Thus, one might argue for ontological emergence by claiming that wholes or systems have causal powers that are emergent from the powers of their parts, such that the system-level causal powers at the macromental level were not determined by the powers of their parts at the microbrain level.[60] With regard to the mind–body relationship, it could be claimed that macrolevel entities, such as mental states, somehow exhibit causal powers not exhibited or determined by their underlying lower-level neuronal states, indicating a real ontological divide.[61] But Searle argues that the causal properties of mental states are identical to the causal

[57] See Corcoran, K. J. "The Trouble with Searle's Biological Naturalism" (2001) *Erkentnis* 55 (3).

[58] Ibid. [59] Van Gulick, n. 45 at 28. [60] Ibid. at 17. [61] Ibid.

properties of brain states. If true, such identical causal properties might be interpreted to provide substantive support for an ontological reduction.

What does Searle's theory mean for mental causation? Mental causation requires that the intentional content of mental states be a causal explanation of my behavior. Searle's claim that the causal properties of mental states are identical to their underlying brain states has an interesting result: any aspect of a mental state that carries causal properties – presumably, their content – carries them via underlying microlevel brain states. The flip side of this is that the aspect of the mental Searle says isn't reducible – phenomenal consciousness – can do no causal work in addition to the mental content as causally realized at the microlevel. That is, Searle's theory indicates that conscious mental states qua their conscious properties – those conscious properties only accessible to the state-holder and not to science – can do no causal work because they are non-reducible. So on Searle's theory, conscious mental states do causal work only insofar as they are related to causal mental content, which he says can be ontologically reduced (at least, Searle has provided no argument that they can't be ontologically reduced). Thus conscious properties, as Searle paints them, are epiphenomenal: the conscious mental state of thirst does not cause me to get a drink – some underlying physical realizer of the causal mental content does. This means that the conscious mental states a judge or jury are looking for when trying to determine if an offender is guilty of a crime are also epiphenomenal: that is, not causally related to the criminal harm the offender was arrested for. In the end, Searle's theory, like Davidson's, does not preserve mental causation and, thus, cannot support criminal verdicts.

The Non-Reductivist Project and Mental Causation

I argued above that two different attempts at a non-reductionist theory – Davidson's, which denies an epistemological reduction, and Searle's, which attempts to deny an ontological one – fail to preserve mental causation. As Tim Crane has noted, in general, non-reductive theories seem to create problems for mental causation because they force higher-level mental properties to compete for causal power with their lower-level physical realizers.[62] Thus, the problem non-reductive physicalism creates

[62] Crane, T. "Mental Causation" in L. Nadel ed., *Encyclopedia of Cognitive Science* (New York: Wiley, 1995).

is the one physicalism was intended to solve. Physicalism was initially posited as a *solution* to the problem of mental causation. As Smart noted in his famous 1959 paper, mental states seem to cause physical effects, but physical effects have complete physical causes. Therefore, Smart claimed, physicalism must be true – to have causal powers, mental states must be simply recognized as physical states.[63]

As we saw earlier, the non-reductivist argues that although mental concepts or entities have physical causal effects via their relationship with the physical entities, these causes and effects cannot be understood in terms of those physical entities via a scientific theory or scientific concepts. As Van Gulick says, "The challenge [for non-reductivists] has been to show that one can 'eat one's pluralist cake' while still remaining robustly physicalist at the ontological level." Specifically, non-reductivists must show that explanations exist at the level of folk psychology – with regard to causal explanations of intentional behavior – that cannot be understood or accessed by use of physical theory.[64] As I noted in an earlier paper, this seems problematic: if the gap posited by the non-reductionist between the mental and physical is ontological, then looking for a particular mental state that caused an action is a fool's errand because there is nothing to be found fitting that description.[65] If, however, the gap is epistemological, then looking for a mental state that caused an action is still a fool's errand because although there might be such a mental state, we are (somehow) necessarily blind to it. In neither case can mental causation, and thus responsibility, be preserved.[66]

Non-Eliminative Reductive Physicalism

Legal scholars Dennis Patterson and Michael Pardo say this about the possibility of reduction of the mental state terms used in law:

> Like neuro-reductionists generally, neurolegalists aspire to reduce the explanation of all human behavior to the level of brain processes. Believing as they do that "the mind is the brain," neurolegalists have attempted to account for mental capacities, abilities, and processes solely at the level of cortical function. As an explanatory account of the nature of human agency, this reductionism aspires to nothing less than the

[63] See Smart, n. 59 and Papineau, n. 17. [64] Van Gulick, n. 45 at 22. [65] Sifferd, n. 41.
[66] Ibid.

> replacement of our talk of beliefs, desires, and intentions with the language of neuroscience.[67]

Pardo and Patterson claim that a reductionist account is going to eliminate and replace our commonsense mental state concepts. But I will argue here the exact opposite is true. A non-eliminative reductionist need not replace and eliminate the folk concepts that underpin responsibility. Instead, reductionist accounts can *vindicate* these concepts and preserve the full-bodied mental causation necessary for responsibility assessments. If a folk psychological mental state is identified with the physical, or the physical is deemed constitutive of the mental, there is no gap we need to overcome to assign the mental causal power. Instead, under non-eliminative reductionism, mental states carry causal properties *just as* physical entities.

Non-eliminative reductionism rests on the notion that a mental state at time *t* is some physical (probably, brain) state. Thus, an agent's desire for a drink of water just is a physical state or process within a particular time slice – no more, no less. This understanding of reduction indicates that all reductions are necessarily non-eliminative. Non-eliminative reductionists do not worry that the mental will disappear or be eliminated via a reduction, just as scientists did not worry that the concept of "water" needed to be eliminated once it was discovered to be H_2O. Even a "bumpy" or "lumpy" reduction does not require a folk concept to be eliminated: only a failed reduction results in elimination. Even assuming physicalism, and thus an ontological reduction of a mental state to some physical state, an epistemic reduction of one theory or framework to another is often fairly rough. As Michael Moore notes, "Even temperature's reduction to mean kinetic energy, for example, was said to have a little lumpiness in it, because of the need to stipulate a randomness of direction in molecular motion in order to reach the identification needed. Lumpier still, it was thought, was the 'mass' in Newton's theory in relation to the mass of which post-relativistic physics speaks; one has to work a bit on what mass means in Newtonian mechanics to get that to be but a special case of Einstein's theories" (see Michael Moore, this volume, Chapter 2).

Of course, if the epistemic reduction proves to be too lumpy – if there is no path to map one set of theoretical entities onto the other – a

[67] Pardo M. and D. Patterson *Minds, Brains, and Law* (Oxford: Oxford University Press, 2013), 28–29.

reduction must be rejected. For example, if marine biology discovered that whales were actually millions of tiny fish stuck together that only appeared to move and live as one organism, it would seem that our concept of whale might be so radically false that we would be required to abandon it. We would say things like: "What we thought were whales turned out to be millions of minion fish." But a bit of lumpiness is to be tolerated, even expected: the concept *whale* was not eliminated when we discovered they aren't fish; instead, we revised our concept of whale to refer to large gray *mammal* that lives in the ocean. In the first case, the existence of whales was undermined by advancing science; in the second case, our folk concept of whale was revised – both vindicated and made more accurate – by advancing science.

The difficulty of a theoretical reduction of mental states to physical states may be exaggerated because for millions of years, subjective experiences of mental states failed to provide evidence of their physical reality. Humans discovered that brains were the locus of mental states long after they knew what mental states felt like, and long after we used mental states to understand and predict our own behavior. However, as Saul Kripke has noted, identity claims need not be known a priori. They can be known a posteriori, or with experience.[68] Human beings did not understand that water was H_20 until we had access to certain scientific tools; similarly, we only came to understand that thirst is some brain state or other via use of certain scientific tools (namely, the tools of contemporary cognitive science and neuroscience). The scientific tools now available to us have allowed us to closely examine the nature of conscious mental states, and it seems the best evidence indicates that the immediate subjective experience of a mental state may be *just what it is like* to have or undergo a particular physical state.[69]

The position of non-eliminative reductionism holds that the causal relationship between folk psychological mental states (such as thirst) and actions (such as getting a drink) can be understood on multiple theoretical levels: for example, at a macrobehavioral level of folk explanations, and at a microlevel, such as the level of functional neuro-connectivity or neurons. Each level may portray a true causal explanation regarding my getting a drink.[70] Understanding the relationship between levels is

[68] Kripke S., *Naming and Necessity* (Cambridge, MA: Harvard University Press, 1980).

[69] Certainly, our current best science has not provided evidence that the subjective feeling of a mental state is anything more than what it is like to undergo that mental state.

[70] Van Gulick, 2001, n. 41.

difficult, complicated work, but several philosophers have already attempted such projects.[71] The non-eliminative reductionist claims this is work that can be done, precisely because there is no unbridgeable gap between mental states couched in folk terms and their physical realizations. The position of non-eliminative reductionism holds that an epistemological reduction of mental states is possible, even though mental states couched in folk terms and brain states enter into fairly distinct systems of laws. Again, it seems very likely this reduction will be bumpy, and that folk concepts may need to be revised, but core folk psychological concepts will not need to be eliminated.

Some have worried that commonsense mental state terms pick out such messy physical states that there may not be relevant physical commonalities that would allow an identity relation. Obviously, a reductionist theory must admit the difficulty of understanding folk psychological concepts in scientific terms, which would be at best an extremely complex process, especially given the breadth and depth of our folk psychological concepts and explanations. One of the reasons some philosophers turned away from type identity theory in the past few decades (and toward non-reductionism) is because of this sort of worry, exemplified by Hillary Putnam's multiple-realization objection. Because pain seems to be realizable by very different physical systems (across different species and persons, and potentially in alien systems) it would seem that pain as a type cannot be identical with some particular type of physical event.[72] That is, it is very difficult to identify a physical commonality between human pain and octopus pain, such that "pain" has a physical identity.

However, it is unclear multiple-realization was a problem worth creating the even deeper problems of non-reductionism (namely, problems for mental causation).[73] There are various ways around the multiple-realization objection. One could argue, for example, that pain is a functional concept, but that local identifications must be made with whatever first-order physical properties or events that instantiate the functional concept within a particular system. In this case mental

[71] See Craver, C. *Explaining the Brain: Mechanisms and the Mosaic Unity of Neuroscience* (Oxford: Oxford University Press, 2007); Kim, J. "Events as Property Exemplification" in *Supervenience and Mind: Selected Philosophical Essays* (Cambridge: Cambridge University Press, 1993); Papineau, n. 17, 1993; and Roskies, A. "Decision-Making and Self-Governing Systems" (2016) *Neuroethics*, 1–13.

[72] Putnam, Hilary *Representation and Reality* (Cambridge, MA: MIT Press, 1988).

[73] Sifferd, n. 41.

properties may be understood as second-order properties, the property of having some further physical properties the instantiation of which plays a certain functional role. Papineau, Lewis, and Kim[74] all use this sort of strategy, arguing that even if multiple realizability means broad type identity claims for psychological kinds such as "pain" aren't possible, local reductions are.[75] According to this view, a psychological kind might reduce to a local neuroscientific (or other scientific) kind. In this case the folk concept of "human pain" refers to a local disjunction of physical states, whereas "octopus pain" refers to a different local disjunction.[76]

Further, as I have argued at various points earlier, non-eliminative reductionists can argue that folk psychological concepts are flexible. The folk may not have realized their folk psychological concepts such as thirst picked out disjunctive local reductions that can be described on many different levels using different theoretical tools. Nonetheless, this appears to be the case, and this realization does not undermine the truth of folk concepts. Instead, these new facts explain why the concepts actually do work to predict and understand human behavior. To take an example from Eliot Sober: it seems true that smoking causes cancer due to some microconfiguration in cigarette smoke. It may also be true that different types of cigarettes have differently-realized cancer causing components at the microlevel. Even so, the macrolevel claim "smoking causes cancer" is true and explains each of the singular instances of cancer caused by cigarettes.[77] An ideally complete explanation of a singular occurrence would include the macrostory, the microstory, and an account of how they are connected. Sober notes that "the kernel of truth in the multiple realizability argument" is that higher-level descriptions may "abstract away" from the physical details that make for differences among the

[74] Papineau, n. 17, 1993, at 25; Kim, n. 71; and Lewis, D. "Reduction of Mind," in Samuel Guttenplan ed., *A Companion to Philosophy of Mind* (Oxford: Blackwell Publishers, 1994), 412–431.

[75] See Funkhouser, E. "Multiple Realizability" (2007) 2 *Philosophy Compass*, 303–315.

[76] Michael Moore also seems to hold this view. This is what he says about the multiple realizability challenge in Chapter 2 of this volume: "[S]urely it will not prove to be the case that some mental state *M* can be realized by some distinct brain state *B* that differs on each occasion that *M* exists. Surely, that is, *M* is realized by a finite disjunction of brain states, B_1, or B_2, or B_3. . . or *Bn*, where n is not infinite and probably not even that large. If so, then *M* will be type-identical to the finite disjunction of brain states, even if it is not type-identical to any one disjunct."

[77] Sober, E. "The Multiple Realizability Argument against Reductionism" (1999) *Philosophy of Science* 66(4), 542–564.

microrealizations that a given higher-level property possesses.[78] However, Sober claims, this does not make higher-level explanations "better" in any absolute sense. Different levels of description have different benefits.

An interesting result of epistemic reductionism is that it can offer the opportunity to pick from multiple levels of true theoretical understandings of human behavior depending on our task. One benefit of folk psychological explanations of behavior is that they are intuitively available to human beings and tied to other important human activities, such as responsibility assessments. As Adina Roskies notes, the multirealizability of folk mental state concepts can be understood to highlight the importance of commonsense psychological explanations, instead of undermine them. Physical states are more finely grained than the mental states they realize, and because folk concepts refer to a disjunction of physical states, the same psychological state can supervene on (somewhat) different physical states across time and persons. Because giving an account of mental causation in physical terms would require reference to a messy disjunction of physical states, it makes sense that human beings tend to use folk psychological terms when understanding and predicting behavior because folk explanations "capture much more succinctly the relevant counterfactuals."[79] To put it another way: explanations of human behavior referencing psychological states express the right level of generality to capture counterfactuals relevant to explanations of human behavior; explanations using brain states do not. This would seem to indicate that the criminal law is correct to use commonsense mental state terms as the primary means to attribute responsibility.

Once we accept both an ontological and a theoretical reduction, the causal properties of mental states become much easier to understand. If my thirst is a physical event or state in my brain, then it holds causal properties identical to this brain state. As some proponents of physicalism noted, human action seems to have a full physical/causal explanation at the microlevel.[80] Even if portions of this physical explanation are indeterministic, the causal chain resulting in an action can be understood

[78] Ibid. at 560.

[79] Roskies, A. "Don't Panic: Self Authorship without Obscure Metaphysics" (2012) 26 *Philosophical Perspectives*, 323–342.

[80] See Smart, n. 47; Papineau, n. 41, 2002; Papineau, D. "Causation Is Macroscopic but Not Irreducible" in S. Gibb, E. J. Lowe, and D. Ingthorsson eds., *Mental Causation and Ontology* (Oxford: Oxford University Press, 2013), 127–151.

as microcausal precursors to that action. Non-eliminative reductionism holds that some subset of those prior microlevel causes of an action – such as getting a drink – *just are my mental states (e.g., my thirst)*. One might imagine being able to look backward, bracket them off, and point at them in awe: "Look, there they are! My conscious mental states! Look at how they caused my behavior!" Or, we might imagine a different scenario where mental states sought after may not be found. For example, in the process of determining if a defendant is criminally responsible, the court might say: "Looking back at the time of the crime, it seems unlikely the defendant had the desire to kill his neighbor. There just no evidence that he possessed this mental state."

Earlier I argued two different versions of non-reductionism fail to preserve mental causation. Davidson's anomalous monism failed to reliably link mental states to the physical states that indubitably carry causal properties. Thus, on Davidson's theory the mental ends up looking as though it is epiphenomenal. Searle claimed that the causal properties of mental states are identical to their underlying brain states, but refused to reduce their conscious phenomenal properties, so conscious mental states qua their conscious properties could do no causal work. Non-eliminative reductionism bites the bullet non-reductivism tries to avoid by claiming that all aspects of mental states, including their causal properties and their phenomenal feel, are ontologically and epistemically reducible. In this case the problem of mental causation fades away. Under non-eliminative reductionism, the mental is a real (physical) thing in the world with physical causal properties. As such, the mental reliably causes behavior.

Conclusion

Criminal responsibility requires the truth of mental causation: a defendant is found criminally responsible for an act where she possesses certain culpable mental states that are causally related to criminal harm. In this chapter, I have attempted to review various theories regarding the relationship between mind and body to identify the theory that is the best candidate to support criminal responsibility assessments by articulating a convincing account of mental causation. Eliminativists can be seen as falling into two categories: Churchland-style eliminativism claims mental states do not exist such that they cannot cause behavior; the new eliminativists denies a type of mental causation they then claim is vital to criminal responsibility. Popular non-reductive theories disconnect the mental from the physical/casual world such that the mental is ephiphenomenal (Searle-style non-reductionism) or can no longer have reliable causal effects (Davidsonian non-reductionism). Thus both types of eliminativism, and non-reductive physicalism, threaten criminal responsibility because they do not provide satisfactory accounts of mental causation.

If we assume the widely accepted positon of ontological physicalism, non-eliminative reductive physicalism about the mind offers the most plausible account of the full-fledged mental causation criminal responsibility requires. This is because it is clear on a non-eliminative reductive account of mind how folk mental state concepts carry causal properties. A non-eliminative reductivist may hold that folk concepts pick out a disjunction of local physical states, and picks them out reliably enough for us to consistently use such concepts to predict and understand human behavior. In any particular case, a mental state has causal properties as a particular token instantiation of this local disjunction.

The criminal law uses folk notions to seek mental states falling within categories or types of mental states within defendants. Criminal courts determine whether specific mental states can be assigned to a defendant and whether these states are causally related to the criminal harm. Such folk concepts seem to refer to conscious mental states, which in turn pick out particular token instances of a local disjunction of physical states. In general, it seems that folk explanations of human behavior are better positioned than scientific explanations to ground responsibility assessments. However, it may be that widespread damage or dysfunction to a particular brain area, connection, or process can be understood to impact an agent's ability to produce or review a particular mental state.

4

Intention as Non-Observational Knowledge: Rescuing Responsibility from the Brain

BEBHINN DONNELLY-LAZAROV

Introduction

Analyses of the relationships between intention, the brain and responsibility commonly proceed on the basis that intention is a state of mind (or that it can be reduced thereto).[1] This chapter will say that it is not (and cannot), and so that the basis for discussion needs fundamentally to be reconsidered.[2] The contribution embarks on such a reconsideration; what kind of thing is intention; how does this thing relate to brain states; and what follows for our concept of responsibility?

Human Action as Responsible Action

States of mind, purposive movements and intentions are the ingredients of human action. If such action is responsible action, it is these ingredients that must somehow make it so. The aim of this section is to explain what makes it the case that each of these dimensions of human agency is integral to an understanding of responsibility. In particular, the brief sketch to follow shows why it might matter at all to that end that intention is not a state of mind.

[1] For a snapshot of the two prominent examples see, M. Bratman, *Faces of Intention: Selected Essays on Intention and Agency* (Cambridge: Cambridge University Press, 1999); D. Davidson, *Essays on Actions and Events* (Oxford: Oxford University Press, 1980). The idea that intention is a state of mind remains dominant, even in fairly radical approaches to action and intention. See, for example, John Hyman, who appears to equate intention with desire; J. Hyman, *Action, Knowledge, and Will* (Oxford: Oxford University Press, 2015).

[2] Setiya sets out the Anscombian origins of the notion that intention is not a state of mind and identifies the thread of renewed interest in that notion. See K. Setiya, *Stanford Encyclopaedia of Philosophy* (entry on Intention) https://plato.stanford.edu/entries/intention/. Accessed on 14 May 2017. This chapter reflects the present author's own particular variant on that core idea.

States of Mind, Movements and Responsibility States of mind (and movements[3]) keep *us* connected to what we *do*. This is not a grand philosophical claim. It observes something simple; we do not *merely* act, as though our actions somehow flow free from our minds and our movements; we act with understanding, through regret, filled with positive emotion, from frustration, recklessly, in hope, for good reasons, calmly, with malice, through purposeful movement. Our mind and our body are present in what we do.

The connection between us and our actions is solidified by the fact that we know it; it is a connection that we experience. Indeed, this knowledge is unavoidable, for states of mind and movement (the stuff of action) *are*, in fact, instances of knowledge. There is nothing mysterious, or philosophically complex, about this insight either. It is just to say that I know I have regret, for example, through regretting, and there is no regretting that somehow stands apart from the knowledge that it is, and so, gives. Moreover, notice the obvious: this knowledge of regret is *my* knowledge of *my* regret and how this deepens the connection further still. When I act with regret, my regret is not only known to me, but it is also inescapably known to me as mine, as my manner of acting. These necessary connections between states of mind and actions are enough to get us to a simple kind of responsibility; I am responsible for my actions because they are *my* actions, known to me as such. That is the experience I have of the things that I do.

Although it is not the subject matter of this chapter, it should be noted that, of course, responsibility is not just a matter of owning our actions in this way and in experiencing the same; responsibility has a moral component too. Here again states of mind are central. It is *particular* states of mind and *particular* purposive movements (in combination) that further determine what moral *form* our responsibility takes, not only whether our actions are *owned by* us but also whether we are culpable, blameworthy, creditworthy and to what degree for what we do. In this way a strategic bomber who knows that his actions will kill innocent civilians and who embarks on

[3] It might seem odd to place movements alongside mind in this regard, phenomena that are in various ways entirely different. The author has written extensively elsewhere about the nonspecial manner in which physical movement is constitutive of action; about how, in being so constitutive, mind and movement are alike. See B. Donnelly-Lazarov, *A Philosophy of Criminal Attempts* (Cambridge: Cambridge University Press, 2015), Chapters 1 and 2; B. Donnelly-Lazarov, 'Intention in Criminal Law: The Challenge from Non-Observational Knowledge' (2017) 30(4) Ratio Juris 451 and B. Donnelly-Lazarov, 'What it Means to be Responsible for Action: Mind and Movement in Criminal Law' (2017) 25 (3) *Journal of Law, Information and Science*.

his mission with glee, *so* acts; his actions are his, comprised of these movements, feelings, impressions, etc., but more than this, his responsibility is, therefrom, given moral substance. The bomber who fears the possibility, knowing it to be unlikely, that his mission might kill innocent civilians and who sets off reluctantly, and full of trepidation, *so* acts, and the form his responsibility takes is likewise informed by these implicated features. This is not to say that one bomber is culpable and the other not (this assessment requires normative analysis); it is to note, merely, that responsibility takes a different moral form depending on the content of those various states of mind manifest in the (necessarily different, therefore) actions.

Intention and Responsibility It is beyond dispute that states of mind and movements are vital for responsibility, connecting us to what we do, constituting those doings, and giving character to the responsibility that arises therefrom. What, if anything, makes intention different? Let's assume, for the moment, no more than the possibility that action characterised by intention (intentional action) gets its nature as such from the knowingness just described; that we act intentionally when we act with or from that conscious apparatus that connects us to what we do. If the assumption seems unwarranted, perhaps in virtue of the oft-paraded and apparently exclusive tie between intention and acting for a *reason*, consider what is missing in 'actions' that are not intentional actions. In 'behaviour' that emerges from sleepwalking,[4] automatism[5] and (true) involuntary[6] states something has gone badly wrong with the knowing, conscious

[4] The status of sleepwalking as a defence in criminal law is interesting; typically it is deemed to fall under the heading of insanity rather than automatism. See, for example, *R v Burgess* [1991] 2 WLR 1206 Court of Appeal.

[5] Sneezing might fall under this head. See *R v Whoolley* (Unreported) Divisional Court of Queen's Bench Division 1997.

[6] Hyman recently has sought to distinguish intentional from voluntary action, denying Anscombe's claim that intentional action is necessarily voluntary. True involuntary 'action' in the view of the present author cannot be intentional. Take Hyman's example of the person who succumbs to rape due to the threat of an even greater harm. There are intentional, voluntary actions that she has done. We might say, for example, that submitting to rape is something the agent does and that she does so voluntarily; she chooses the least dreadful option for the reason that she wishes to avoid the more dreadful one; this is the classical form that decision making, and action, takes. What is not voluntary in this scenario is the sexual intercourse that ensues. In this regard, the woman becomes an object not a participant. She is not engaged in sexual intercourse; it is done to her and it is neither intentional nor voluntary for it is not an action of hers at all. (For this reason, the offence of rape is undoubtedly present. Submitting to rape is a form of absence of consent to sexual intercourse; it is one way in which rape might occur.) See Hyman, 2015, n. 1, Chapter 4, 'Voluntariness and Choice'.

framework to human endeavour.[7] And, observing the behaviour, we tend to take the view that the person is not really acting at all; that the characteristic intentional quality of human action is missing. The best explanation for our response is not that the agent lacks reasons but that the agent lacks the right kind of knowledge. Human action *is* intentional action in virtue of the knowingness that characterises (and does not stand outside) its constitutive mental and physical capacities, a knowingness that is absent in instances of involuntariness, automatism and the like.[8] So, the sleepwalker, in the moment, might have clear reasons for her action, might move very effectively, might be frustrated or happy. But, what she lacks is the knowledge that these dimensions of her action necessarily give to *her* (and consist in) when she acts intentionally. In sleepwalking, it is quite true to say that 'she' really does not know what she is doing.

What does it mean for knowledge of this sort to be *present* in a given *action*? Consider a person who walks to work slowly and reluctantly. We can say that this person is doing the action of [walking to work slowly and reluctantly] if these features of the behaviour are known to him *as* partially definitive of the thing done; if he might consider the question 'Why are you walking to work slowly and reluctantly?'[9] to be one that does indeed address an action of his. If the reluctant worker is completely unaware of his manner of walking and feeling, if these are background influences buried in his psyche and merely imposing themselves on his

7 Hornsby's rejection of causal accounts of action might have led here too, but she is perhaps equivocal on the point; 'When she is doing anything, thanks to her capacity to do it, there will be explanations of what she is doing which show her as having a reason to be doing it. She is a rational creature, leading a life, equipped with powers of thought and self-movement'. Hornsby's observation is a good one save for the explanatory prominence given to reason. A narrow, purposive account of reason does not *give* the action (although it can allow us to see that there is one). Rather, any of the constitutive elements of action can legitimately feature in a suitably broad explanation for action; these elements will include not only reasons for action but also movements, thoughts, life experiences, emotions, memories etc and all of this, alike. See J. Hornsby, 'Causality and the Mental' (2015) 29 *Humana.Mente Journal of Philosophical Studies*, Vol. 29, 125–140. For an elaboration of the idea see Donnelly-Lazarov, 2015, ibid.

8 In involuntariness, note that the knowingness that is present is precisely that of being acted upon, not acting.

9 Famously the applicability of Anscombe's 'Why' question allowed, for her, the identification of intentional actions. See G. E. M. Anscombe, *Intention* (Oxford: Blackwell, 1957). But, this particular 'why' question is not quite Anscombe's. In putting the question in the manner here put, the explanatory dimension to human action is brought into play. For this reason we are far more likely to be able to say not only that the man walks to work but that he does so slowly. This reflects a broader account of intentional action than those accounts that commonly follow Anscombe and typically orient action (exclusively) around reason in its motivational, teleological or purposive sense.

action, it cannot then be said that walking slowly or reluctantly to work is something that intentionally he does. And, of course, the matter may be rather more prosaic. The same might be true of walking slowly on ants; it can be an intentional action, or not, depending on what is known *as* the thing done. What matters when it comes to defining the contours of intentional action is the recognition that accompanies it.

This is an account of intention that most will happily reject. It is introduced briefly at this juncture because here the importance of intention in understanding the origins of responsibility needs to be set out, and it makes little sense to use an account that the author does not share. Moreover, at this point, it is the positioning of intention that matters, and the approach mentioned chimes with pretty much every other insofar as situates intention somewhere between mind and act, and there is certainly something about intention that places it there; between our conscious states and the responsible actions that flow therefrom.[10] Indeed, this positioning is precisely why intention matters uniquely for responsibility. When we act, something happens to states of mind and movement. They are no longer merely states, or merely movements, they are the *capacities* that take us to action; present in what we do, as much as in our minds, our brains, or our bodies. If we are to know how our potential to act with and from states of mind is realised in responsible human action, it seems, then, it is to intention that we must turn. If this is an explanatory burden that underlies *any* account of responsibility[11] what changes in the account here proposed is the characterisation that will be given to intention. In short, it cannot be the case that intention plays this transformative role, allowing states of mind and movements to instantiate themselves in human actions, while at the same time, being, itself, but one of these same states of mind.[12] To understand the role intention does play in responsible human action is properly to characterise what it is.

Brains and Responsibility The place of the brain in this story is not in all respects obscure. Certainly, it is beyond dispute that some important

[10] Indeed, this admitted positioning of intention might give us cause to be suspicious of those analyses that at the same time present intention to be a state of mind like any other.

[11] See, for example, Michael Moore's contribution to this volume, where he explains an account of intention in a functional sense.

[12] Searle asks similar questions: 'But if we say that intentions cause actions, then what happens to beliefs and desires? Can they still be causes too? And if so, how does their causal role relate to that of intentions'? What is an action anyhow?' J. Searle 'The Intentionality of Intention and Action' (1980) 4 *Cognitive Science*, 47–70, 48.

and well-defined relationships hold between states of mind and the brain. It is further beyond dispute, of course, that big questions remain; about the metaphysical relationships between the mind, the brain and responsibility. In this regard the deep waters of reductionism are unavoidable. This is not the place to address the important relationships or to engage in the reductionist debates. Michael Moore's contribution to this text[13] navigates the field and the reductionist waters comprehensively, and other earlier chapters address them too, setting out the nuanced commitments involved and the implications of each. Here it suffices to point out that it is obvious why the reductionist debates matter for responsibility. They matter precisely because of the features of human action outlined earlier for in virtue of them we experience ourselves to be responsible actors; we know our actions to be *our* actions. And, now, the idea appears troublesome indeed that we might, despite our faith in responsibility, be 'nothing but a pack of neurons'.[14]

From this brief sketch the following might be mooted. We cannot hope to understand the origins of responsible human action unless we understand how states of mind connect us to what we do. And we cannot understand that connection unless we understand the apparently transformative role of intention. To determine whether responsibility is undermined or challenged, in any way, by a neuronal context is so to be quite precise about what connects, not only states of mind, but also intention, specifically, to the brains that we have.

The State of Mind Thesis; Any Basis for Challenge?

The preceding section showed how it might matter to responsibility that intention is not a state of mind, but it remains to be seen whether the, dominant view that it *is* such a state can seriously be challenged. According to the traditional position, intention is a multifaceted phenomenon; it takes three forms: the intentions we have that precede action (and may not be acted upon at all); our intentional actions themselves; and the intentions with which we do those actions.[15] Because, despite these three forms, intention is nonetheless conceived to be a single phenomenon, it follows that intention in *each* of these manifestations is

[13] See this volume, Chapter 2. [14] See ibid.
[15] See Anscombe (1957), n. 4, where the three faces of intention are clearly set out.

a state of mind.[16] For present purposes, we can narrow the matter somewhat. Our concern is with the relationship of intention to responsibility, and in this regard, the application of the state of mind thesis to intentional actions is what matters most; it is these, centrally, that human beings are responsible *for*; not for those intentions that apparently precede action but for intentions that somehow make their way into the active world in the form of intentional actions. Naturally, reductionism's appeal, and the enquiries that occupy it, are most prescient here; the central question becomes this one: what is left of responsibility if the intentional dimension of *intentional actions*, the things that we are responsible for, is reducible to brain states?

Where, if at all, might we find reason to challenge the idea (and the ensuing appeal of reductionism) that intention is a state of mind and, so, that intentional actions implicate that intending state of mind? Consider intentional actions themselves. A central implication of the state of mind thesis is that these intentional actions are *both* [intentional] and [actions]. This is to say that there are two dimensions of the phenomenon and that these are in different domains; the nature of intentional actions as intentional is of the mind; their nature as actions is of the world. The domains need each other (there cannot be an intentional action without both an intentional state of mind and the action that instantiates it), but they are discrete. Moreover, intention is but one state of mind; there is nothing distinctive about it or about the way in which it feeds into actions; it occupies the same role in our minds, and in the actions that flow therefrom, as recklessness, suspicion, knowledge, happiness, caution, remorse, regret, attentiveness, or naivete.

The import of this bifurcation can, perhaps, best be explained through the lens of criminal law, an enterprise that institutionalises the separation and identifies its apparent utility for understanding human responsibility. In that institution, any criminal offender is responsible not for what he does but for the fact that he has [done it] with a [particular state of mind], and so if we are to understand whether a defendant is culpable in the relevant way we must understand two things: what he does and the state of mind he has in doing it. This is to say that a defendant whose car careers off course due to severe unexpected winds, colliding with and causing the death of a pedestrian, has not committed the offence of

[16] See Setiya, n. 2 for an account of various attempts to unify the three manifestations of intention.

murder. So we might *say* that this defendant kills the pedestrian but concede that it is not his purpose to kill or to cause serious harm. And, so, if he has done the action of killing, still he has not done it with the state of mind designated in the offence of *murder*. Murder requires that the defendant, in its terms, 'intends' to cause death (or really serious harm), and this is what distinguishes the particular action of 'murder' from involuntary manslaughter, for example, or (as here) from causing death, lacking any culpable state of mind.

This familiar paradigm of actus reus and mens rea is one that admits of an array of interesting legal philosophical questions, none of them easy and many replicated in the reductionist debates just alluded to. How is it that the mind is related to the action; can it be, and if so how can it be, that the mind causes the action; is it the action or the mind or both that causes the *result* of the action; is the action confined to our bodily movements or does it include their effects; how can the institution defend culpability if intentions and other relevant states of mind are a product of, or reducible to, brain states; how can law distinguish morally between one such state and another? But all these questions assume the two (aforementioned) truths about human action. First, crimes involve both states of mind and actions as *distinct* elements; mens rea and actus reus.[17] Second, intention is but one such state of mind.

There are four reasons, for present purposes, to seek an alternative concept of intentional action. First, there is good cause, at a very basic, experiential, level, to reject the view that intentions are states of mind like any other. Our experience of intention is not like our experience of states of mind. Indeed it has often been observed that although there is something that it feels like to be happy, to be thinking, to be resolute, to move, to be carefree, it is more difficult to put a feeling on intention. Is there anything that it feels like to be intending? Can we point to an example of when we felt it? For how long did we have the feeling or the intention? How many times did we experience it today?[18] These questions do not readily admit of sensible answers and so it seems as though there is something different about intention, something that

[17] Moore is right to observe that analyses of this aspect of the mens rea/actus reus paradigm do not typically bear upon debates about dualism. Usually, attempts to press the point are bound to fail. See this volume, Chapter 2.

[18] Teichman makes similar observations. See R. Teichman, *The Philosophy of Elizabeth Anscombe* (Oxford: Oxford University Press, 2008), 2.2.

distinguishes it from states of mind. Now the possibility is at least worth exploring that if intention *is* experienced at all, then it might be the thing we experience it to be, and we do not experience it to be a state of mind.

Second, our experience of action tends to confirm the view that intentional action is not a dual phenomenon. We readily recognise when we are frustrated, angry, happy, intoxicated, perplexed, amused, tired, resolute, committed or relaxed. We can also point to instances of reflection, reasoning, thinking, knowing, hoping, anticipating. Some of these phenomena are associated with emotion, others with intellect, still others with a combination of the two; some may be basic states, while others implicate basic states. Still there is something clearly *of* the mind in all of this; these are our knowing mental states or instances of them. Now, when we act, any or all of this apparatus can be constitutive of the thing done; this is just to say that I might no longer just feel hesitant, uncommitted and lethargic about my professional commitments but that now [I go to work slowly and reluctantly]. This is what I do, and my action is not a mere collection of still passive states of mind and movement. It is actually done *as such*; it *is* comprised of these states in their interacting active form; there is nothing else for it to be. Now this being the case, it no longer makes sense to conceive of actions as a [thing done] and states of mind/movement as standing in some relation to that thing. For, there is nothing that the thing done can be other than these states in their active form. And this is how we experience intentional action; as a perfectly unified thing, not as some ontologically peculiar entity comprised of itself and states of mind that somehow feed into it but are not it.

So, our experience of agency gives us good ground to support the notions (a) that intention is not a state of mind and (b) that we act intentionally when passive states are no longer just states; when they are the active capacities that our actions are constituted from. But this is to assume the explanatory burden for it might be said that the present author merely presupposes it to be intentional action that is constituted therefrom. It remains to be explained why this might be so. Why is it that these actions, implicating the entire reach of human capacities, can all be described under the heading of intention? What does this tell us about the character of intention?

These questions require an intellectual response that experience cannot provide. Before attempting to offer one, it is worth noting a third reason to challenge the state of mind thesis. This is the challenge from frustration and the hope that it might be overcome. The reductionist debates are complex and do not admit of a clear way through. In their particular, antireductionist, debunking of both dualism and reductionism, Patterson and Pardo have gone some way to seeing a way out. By their Aristotelian understanding, 'to have a mind is to possess an array of rational and emotional powers, capacities, and abilities exhibited in thought, feeling, and action'.[19] It follows that it is a false errand to seek this mind in some place for powers, capacities, abilities and the like are not substances at all, physical or otherwise. Moreover, if capacities are to be identified with their holder, then it is the person as a whole and not the brain that does the holding.

Patterson and Pardo do not aim to provide an explanation for our ownership of these capacities/this mind; for how it is that a person with a brain, and other attributes, might have a set of capacities such as these. If it is not their purpose to provide one, the problem survives, nonetheless, that it is just not obvious how an explanation might come about. One might accept entirely their understanding of the mind as an array of powers, capacities, abilities, but there is still a need to understand how it is that an entity such as us can have an idea, say, and the experience of an idea. We understand why, given its makeup, a car has the capacity to drive, but we cannot readily understand how, given our constitution, a person has the capacity to think, feel, know, even to move purposefully towards a desired end.[20]

These observations might seem completely irrelevant to the matter at issue or to an account of intention such as the one here mooted. Why do they give us any reason to look for a new formula? Patterson and Pardo, via their opposition to reductionism, provide a convincing account of the mind. But, should it turn out that intention specifically is not a state of mind, and that there is something entirely other than the mind that demands explanation, the need falls into the background to build an account of intention around reductionism or opposition to it. Moreover, depending on what this account has to say, it may have implications more generally for how we might understand the relations between their capacities and the human beings whose capacities these are. So, the

[19] M. Pardo and D. Patterson, *Minds, Brains, and Law* (Oxford: Oxford University Press, 2013).

[20] Many will dispute the idea that the capacities-thesis does not count as an explanation. Still, it seems very likely that our explanations are only good by a measure that substantially modifies and blurs our concept of causation, or similarly broadens our understanding of what it means to understand or to provide an explanation. Either way, the explanatory reality is one of failure.

possibility of another paradigm provides an inquisitive motivation for asking whether intention really is a state of mind.

The final reason to seek an alternative is an intellectual one, and it is to this that the remainder of this chapter will refer. It just seems increasingly unlikely that intention is, after all, a state of mind or that intentional action is the two-dimensional mind/act entity that many have assumed it to be.

Reconceiving Intention: Non-Observational Knowledge in Acting

What is intention? For the sake of continuity, the institution of criminal law can again provide focus. It allows us to isolate intentional actions specifically. The practice is not interested, typically, in those defendants who intend to kill but don't act on that intention. It is interested in intentional actions themselves – where the practice can properly attribute responsibility, in the form of culpability, to a defendant. So, let's start here. What is intentional about intentional actions?

In proceeding, it is useful to forget for a moment law's mens rea/actus reus paradigm. Consider, instead, its offences. Theft, in law, is the [dishonest appropriation of property belonging to another with the intention of permanently depriving the other of it]. If a defendant is found guilty of the offence of theft, we can say that theft is something he *did*; it was an action of his. And, of course, to say that it is an action of his is to say that what he did amounted to [the dishonest appropriation of property belonging to another with the intention of permanently depriving the other of it]. A central thesis of this chapter is that *this* is what intentional actions look like. We have capacities [physical and mental] that, alike, are brought to action. Because these capacities are known in their performance and known as ours, the actions that are constituted through them are properly considered to be our intentional actions. Here, there is no distinction between mind and action; there is just intentional action; our mental capacities are constitutive of these actions just as our physical capacities are. How might these intentional actions be known? If non-observational knowledge might not be the obvious place to start in answering that question, an exploration of the kind of thing it *is* suggests some promise.

A comparison with our capacity to see can help set the scene. How do you know that you see? How do you know that you are seeing this chapter; the room you are in, its colours and decor? In asking the question, the concern is not with how you know that it is this chapter or this room that you see, it is rather with how you know that you *see* these things. To answer that question, consider what you do not need to do. You certainly do not need to reflect on the matter. You do not need to ask yourself; gosh, how is it that I know that I am seeing the chapter? Nor do you need to stand aside from your seeing-self and observe that process, as though you might need to 'see', or otherwise somehow to behold, the seeing in order to know that you see. No process of second-order observation is required. Indeed, perhaps, the question, 'How do you know that you see the chapter?' just seems a little odd; it might so seem because 'seeing the chapter' *is* knowing that you see the chapter. To neutralise a metaphor, seeing really is knowing. Indeed, we have already noted that in this regard so are all our senses/emotions/cognitions/purposive movements knowing ones, knowing, that is, in and of themselves.[21]

Here, I want to defend the claim that acting, like seeing and these other states, *is* knowing. [Carefully breaking eggs] is knowing that I am carefully breaking eggs. Now, at first glance it might seem that any such comparison is misplaced. I know *that* I *see* a tree because *actually* seeing a tree tells me the same; this is just what seeing is like. Moreover, because seeing is like *this*, an independent concept for this particular kind of knowing is superfluous. Things are a little different when it comes to the knowledge that acting *is*. True, I know that I am carefully breaking eggs in carefully breaking eggs, but we do have an independent concept for the knowledge that doing is. That concept is intention; to know that I am carefully breaking eggs *is* to so break eggs *intentionally*. The need for an independent concept arises for this reason; in acting, we exercise our capacities to *do*, and what we know, therein, is practical. But the action is not known as a set of atomistic, discrete capacities, it is known as such. And, its nature as such varies depending on what is, in the moment (and depending on one's purposes, etc.), in fact, known. This is to say that the

[21] For some discussion of the same idea, see B. Donnelly-Lazarov, 'Intention in Criminal Law: The Challenge from Non-observational Knowledge', (2017) *Ratio Juris*. Searle employs the same analogy to explain non-observational knowledge: 'And, again, the parallel with perception holds. Just as at any point in a man's conscious life he knows the answer to the question, "What are you doing now?" so he knows the answer to the question, "What do you see now?" In both cases the knowledge in question is simply an awareness of the conditions of satisfaction of a certain sort of presentation'. (See Searle, 1980, n. 12, 56. Searle's account is not comparable to the one here proposed because of his insistence on the existence both of discrete prior intentions and such things as physical actions.

thing known gives life to the thing done as the thing done gives life to the thing known. This is what practical knowledge is like; the evident (causal) circularity produces no conundrum for it is a characteristic of the phenomenon at issue.

However complex this might appear, it suffices to say, for now, that the initial comparison stands. I do not need to observe my 'seeing' to know what I see. Equally, I do not need to observe my 'doing' to know what I do. In both cases, my knowledge is non-observational. Knowledge of action is distinctive because it is *practical* non-observational knowledge, and intention is the name we give to it.

Intention as Non-Observational Knowledge

Non-observational knowledge is not a simple idea.[22] The present author has previously suggested that it can best be grasped via comparison with observational or theoretical knowledge. In knowledge of this sort our capacity to know is directed to objects. I know that it is raining when I feel a few spots of water on my head, by observing the clouds and the rain that falls from them, when I hear the sound of water bouncing off the ground, by questioning my friends about the meaning of the precipitation I feel. Crucially, it would be very peculiar indeed to claim that this knowledge *consists in/of* the water bouncing off the ground, etc.[23]

Perhaps surprisingly, this peculiar claim is precisely what makes sense of knowing what I do; my knowing does indeed consist in my doing. Take a rather complex action like playing chess, one that involves the exercise of many complex mental capacities as well as simple physical ones. How is it that I know, from the exercise of all of this, that it amounts to playing chess, that this is what I am doing? Imagine that the answer to this question is that knowing what we do requires us to exercise a capacity to know. It would seem that alongside our actions, we would need to perform another set of observational actions aimed at keeping us informed. So, although it must be true that the chess player knows what she is up to, this cannot mean that she must constantly

[22] Some take the view that the fault is Anscombe's. See Schwenkler, for example, who says that her account: 'is anything but systematic, and she appeals to the concept of practical knowledge with a bewildering array of formulations, several of which are hard to see as connected to the others'. J. Schwenkler, 'Understanding "Practical Knowledge"' (2015) 15 *Philosophers Imprint* 15.

[23] See above at n. 21.

attend, reflectively, to her movements, processes of reasoning, motivations, emotions, plans and to the entire panoply of physical and mental processes that keep her action going. Nor can it mean that she must then put all this together, make sense of it as a unified activity and conclude that she is playing chess. It is hard to see how anything but cognitive inertia and complete inaction would be the outcome of any such parallel process.[24] This is not, of course, to say that we never have good grounds to reflect on what we do; it is to say rather that this is not how we come to know what we are doing. Reflection on action is reflection *on* what I know already, in and from action itself. My carefully breaking eggs *is* my knowing that I am carefully breaking eggs.

At this juncture it is necessary to make the connection plain between non-observational knowledge of action and intention. Those familiar with the work of Elizabeth Anscombe will recognise her legacy herein. Intending, for Anscombe, *is* our non-observational knowledge of being on the way to intentional action.[25] At first sight, this seems to leave something out; indeed, one reason why her work has, at various times, lost the impact it deserves is Anscombe's apparent failure to account for pure intention, the intention I am supposed to have prior to acting. Why might her position seem susceptible to the objection? Let's assume it to be true that we know our intentional actions non-observationally and that non-observational knowledge is the mark of intentional action. What Anscombe accounts for then, it seems, is intention in one of its guises – intentional action. Indeed, by her account intention has, as yet, no reality at all outside the action that gives it life. Here is the problem: I can intend *now* to make pancakes at the weekend, and this intention is entirely apart from the actual making of pancakes. So, even if making pancakes is somehow known to me non-observationally, and even if this is somehow connected to my action as an intentional one, what of it? This tells us nothing about *pure intention*, which is wholly prior to and

[24] O'Shaughnessy makes a similar point: 'As I attempt to relate observationally to my own actions, either observation seems merely to skid over the surface of the act or else the act begins to come to a standstill'. B. O'Shaughnessy, *The Will* (Cambridge: Cambridge University Press, 2008), 332.

[25] It had often been assumed that this account of intention and intentional action was merely a stepping-stone, one that contributed to the far more sophisticated theory provided by Donald Davidson. Thankfully, this account of events is now doubted by many, among others; Falvey, Moran, Thompson, Rodriguez-Blanco, Setiya, Paul, Hursthouse and Hornsby.

apart from the action that might fulfil it. Perhaps, then, *this* pure intention belongs, after all, to the mind.

The objection is taken to be a fatal one by those who fail to embrace the radical nature of Anscombe's account. But Anscombe does not omit to account for pure intention; part of her aim is to demonstrate that it is not 'there' to be accounted for. To defend this position, Anscombe must show at least two things: first that we do know our intentional actions non-observationally, and second that this knowledge completely accounts for the nature of intention. Donald Davidson denies both claims. In denying the first he provides an apparent counterexample designed to show that we can act intentionally without knowing what it is that we are doing. It goes as follows: a man who wishes to make ten carbon copies of a print, knowing only that he is trying to do so, and believing that he is unlikely to succeed, may actually achieve his end. According to Davidson, this agent has intentionally made the ten copies even where he lacks knowledge of his success (of the non-observational or any other kind).[26] In addition to Davidson's, there is another more straightforward objection to the idea that our knowledge of action is non-observational. Consider the fact that if I wish to paint a wall yellow, I must recognise, observationally, that the paint is yellow. Indeed, I must know too that it is paint, that there is a wall, etc. And, if walking slowly on ants is to be something I intentionally do, I must know, observationally, that the ants are there beneath my feet. So there are two possible strong objections to the notion of non-observational knowledge of action. First, I can do intentionally what I do not know at all. Second, I can do intentionally what I know only in virtue of observational processes.

First, to Davidson's objection: Davidson misunderstands non-observational knowledge, and his critique is poorly aimed. Crucially, for Anscombe, non-observational knowledge is not some capacity we have that allows us to know what we do; acting, like seeing, *is* knowing. Moreover, because this quality *characterises* intentional action, if the knowledge is absent, then so too is the action. If this is the case, there must be something amiss in Davidson's apparent counterexample. What can it be? Consider some of the descriptions we might give of the action he purports to describe. Might we not say of the actor that he [makes copies, believing he has little chance of

[26] D. Davidson, 'Actions, Reasons, and Causes' (1963) reprinted in *Essays on Actions and Events* (Oxford: Oxford University Press, 1980), 3–20.

succeeding in making ten, but hoping he will]; that this is something he does? Moreover, might he not recognise his action so? Another way of putting the point is to consider the answer to Anscombe's 'why?' question.[27] Why is the actor pressing down on the paper? 'To try to make ten copies' is a response he might give. In fact, [to make ten copies] is a response he is very unlikely to give. The [trying] response is more plausible in the sense that 'trying' is particularly relevant to this actor's manner of acting; he lacks faith in his ability to succeed but acts nonetheless.[28]

How are these alternative accounts and descriptions of the thing done relevant to the question of non-observational knowledge? Davidson's agent, in trying, knows his trying; in trying with uncertainty he knows this too; in achieving an unlikely result through luck, he will come to know this; all of this is constitutive of the thing done, and all of it is known, not atomistically, but through and as the action done. So, sure, we can *say* here (only ex post and from an external perspective) that the agent actually [made ten copies] and propose that this is an intentional action of his. But, the important question is why would we say such a thing? We might do so either because we are a philosopher seeking to understand the action in a curiously unilluminating form, or because we are just not the agent; his non-observational knowledge is precisely what we cannot have.[29] If we ask the agent, he will not tell us, 'I am making ten carbon copies'. More likely he will report, 'I am trying to make 10 carbon copies', or 'I hope to make 10 carbon copies'.[30]

The second objection I mentioned earlier misunderstands the proper role in action of *observational* knowledge. Of course when I paint a wall yellow I need to know that the paint is yellow. But, in a point made well by Falvey, this enabling knowledge is part of the activity itself. This is to

[27] See n. 9 above for a reference to 'why' questions. Anscombe's 'why' question is designed to determine whether, and what, action is intentionally done. The answer to the question gives the 'what', the applicability of the question gives the 'whether'.

[28] The present author has examined the matter elsewhere. See Donnelly-Lazarov, n. 3, 2015.

[29] Thompson makes the point: 'You can say he made ten copies intentionally if you like, but it will not be an illustration of the topic of Anscombe's book ...' M. Thompson, 'Anscombe's Intention and Practical Knowledge' in A. Ford, J. Hornsby, and F. Stoutland (eds.), *Essays on Anscombe's Intention* (Cambridge, MA: Harvard University Press, 2011).

[30] See, for an elaboration, Donnelly-Lazarov, n. 3, 2015, Chapters 1 and 2.

say that I need to *know* the paint is yellow *in order to* paint the wall yellow.[31] Just in the same way, I need to have a *reason* to paint the wall yellow; I need an *understanding* of the processes involved; I need to *move* my arm in the appropriate way; I may need to *remember* where my brush is; I need to *communicate* with the paint shop owner. I need a range of capacities in order to perform my task, but still, I know that I am doing it without any exercise of any such capacity. I know it non-observationally.

How does any of this help with the second burden that Anscombe must discharge; to show that non-observational knowledge of action exhausts the categories of intention; that 'pure intention' is a myth? Consider the statement: 'I intend to make pancakes at the weekend'. There is nothing erroneous about such a statement in Anscombe's account and it does, entirely properly, express an intention for the future. But, for Anscombe, this agent is already on the way to the intentional action of making pancakes, and so, action has already, albeit in a very embryonic form, begun. Intention has no reality that is apart from the non-observational knowledge that (only) intentional *action* instantiates. It does not have a pure form.

This aspect of Anscombe's position will remain unattractive and surprising to many. For her, when I say, 'I intend to make pancakes', I express a conative perspective, one that consists *in* action, in *being on the way* to making pancakes. But many have pointed out that this apparent peculiarity actually reflects our ordinary manner of speaking. Falvey refers to it as the 'openness of the progressive', using the following example: 'Suppose a friend stops by my house and wants to go for a walk, and I say, "I can't; I'm making bread". This could be true even if as I say it I'm sitting on the couch reading the newspaper – perhaps I'm waiting for the bread to rise before putting it in the oven'.[32] Anscombe's claim that intention is being on the way to intentional action reflects the good sense in this form of expression. Intention is our non-observational knowledge of being on the way to intentional action; the 'being on the way' might be action in a very preparatory sense only, or it might be close to perfecting a particular thing to be done.

[31] See K. Falvey, 'Knowledge in Intention' (2000) 99 *Philosophical Studies* 21–44 at 24. See, for an alternative analysis, M. Thompson, *Life and Action* (Cambridge, MA: Harvard University Press, 2008) 140–1.

[32] Falvey, ibid. at 22.

A Departure from the Brain?

Hursthouse has commented: 'It is the beauty of Anscombe's description of practical knowledge that ... it makes my so-called knowledge of my intention in acting extend beyond my head into the world, without undercutting its essentially first person nature'.[33]

Hursthouse's point is precisely the one that neuroscientists might attend to. The idea has been dominant that if we are to understand the relationship between states of mind and responsibility, we must assess whether these states can be reduced to the brain; that reductionism and counters to it are a vitally important part of the landscape. Intention, then, is but one such state of mind. This idea can be challenged if the account of intention introduced earlier has any ring of truth to it; i.e., if intention is *of* action and not *of* mind. When we act intentionally passive states of mind and body are no longer passive, and they are no longer known in their atomistic form. What *is* known (non-observationally) is action; a composite entity. And, we *do act* intentionally just when non-observationally we know the same.[34] The intention gives the action and the action gives the intention. There is no place for the mind in this story; being on the way to playing chess is not *in* the mind. Now the challenge to reductionism can simply be set out. If intention is not a state of mind, it now seems unlikely that we might unearth an intention 'in' the brain that is not (even) in the mind; minimally we will find it much harder to go about looking for it.[35]

[33] R. Hursthouse, 'Intention' (2000) 46 *Royal Institute of Philosophy Supplement*, 83–105. The paper was first published in the early 1980s, and Hursthouse in 2014 affirmed her commitment to it.

[34] So, the knowledge is both caused by and causes the action. It is only in this (not 'efficient') sense that non-observational, practical knowledge, and so intention, is causal. Moran explains this effectively: 'If I do not have non-observational, practical knowledge that I am clicking out a particular rhythm with the pump then as described in this way that does not count as an intentional action of mine ... This leads to a particular sense of practical knowledge as being the cause of what it understands'. R. Moran, 'Anscombe on "Practical Knowledge"', in J. Hyman and H. Steward (eds.), *Agency and Action* (Cambridge: Cambridge University Press, 2004), 43–68. Moran also usefully here observes that non-observational, practical knowledge is to have a causal constitutive role, not an efficient causal role.

[35] Moreover, because intention gives active life to otherwise, discrete, passive states of mind and body, a revised account of its nature might have significant implications for our understanding of states of mind and their relation to the brain, too. Consider very recent work on the legally vital distinction between knowledge and recklessness. There the researchers purported to identify the brain responses associated with these two mental states. However, note that it was not merely mental states that they wished to observe but

How might the recharacterisation of intention and the rejection of a reductionist paradigm bear upon our concept of responsibility? Importantly, we need not now assume that the challenge to responsibility that reductionism presents remains a live one; for if intention is not a state of mind, and if neuroscience is to illuminate our understanding of intention, then the vital consideration is how brains relate to the *practical, non-observational* knowledge that gives life to and persists in what we do, not in mental *states* that we 'have', or 'are in'. But now too the challenge to reductionism itself comes into view that practical, non-observational knowledge leaves the brain behind in a way that passive states of mind do not. If we want to understand intention, the real challenge is to understand how our capacities to act, founded on our states of mind and our movements (whatever *their* associations with the brain), become realised in actions of ours, known to us, non-observationally as such; actions for which we are responsible.[36] How does that transformation occur? This chapter has suggested one possible answer – that the transformation occurs just when non-observationally we know the same. Whatever the answer, the plausibility of the question predicts the likelihood that a lot of what intrigues us about human action is not readily amenable to neuroscience, however sophisticated, however precise.

Something more needs to be said at this point about why precisely it is that the brain loses its traction if intention is so conceived. A number of negatives can be asserted about intention. Intention by this account is

Not a state of mind
Not a 'thing' and, so, not some*thing* we have

mental states, active, in action (knowing or reckless action, depending on what the participants were told). Now the problem to emerge is that observing *this* does not amount to observing mental states at all. When we act, our states are precisely (and normatively most relevantly) *no longer* states. (See, I. Vilares, M. Wesley, W.-Y. Ahn, et al. 'Predicting the Knowledge-Recklessness Distinction in the Human Brain' (February 9, 2017) 114 *Proceedings of the National Academy of Sciences* 12, 2017)

36 It might be said, at this point, that it is precisely to the brain that we must turn if we are to understand this non-observational knowledge. But, such a response might be deemed obtuse; why ignore the clearly identified phenomenon at issue? Searle makes an interesting observation about the peculiar nature of explanations that seek to define any object by means that avoid the very thing encountered: 'If one were doing a study of hands or kidneys or of the heart, one would simply assume the existence of the entities in question, and then get on with the study of their structure and function. No one would think of saying, for example: "Having a hand is just being disposed to certain sorts of behaviour such as grasping . . ."' J. Searle, *Intentionality: An Essay in the Philosophy of Mind* (Cambridge: Cambridge University Press, 1983), 263.

Not a capacity
Not present in a pure form
Not reducible to the capacities it activates[37]
Not a matter of efficient causation

A phenomenon of this sort seems incompatible with any form of reductionism. How, for example, can we reduce to the brain something that human beings do not have? Why would we consider the reductionist enterprise, or counters to it, to carry any great explanatory worth in this regard? The positive features of intention seem to bolster the case. Intention is

Acting as knowing
Non-observational, practical knowledge
Recognition of what we do
Just knowing
Active
What causes my actions to be *actions*
Caused by my actions being actions
Transformative
Descriptive of the human manner of acting.

Given these features of intention, what might knowledge of the brain tell us about intention? Can non-observational knowledge be observed? Can being on the way to making pancakes be observed in the brain? These are important questions for philosophers and neuroscientists alike.

[37] It is noticeable that studies on intention consistently fail to observe any distinction between decisions and intentions. See, for example, the well-known study: B. Libet, C. A. Gleason, E. W. Wright, and D. K. Pearl, 'Time of Conscious Intention to Act in Relation to Onset of Cerebral Activity (readiness-potential): The Unconscious Initiation of a Freely Voluntary Act' (1983) 106 *Brain*, 623–642. For more recent work, see: C. S. Soon et al., 'Predicting Free Choices for Abstract Intentions' (2013) 110(15) *Proceedings of the National Academy of Sciences* 6217–6222.

5

Efficient Causation and Neuroscientific Explanations of Criminal Action

NICK J. DAVIS*

Introduction

In recent years, legal scholars have looked to the brain and behavioural sciences to help in delimiting criminal behaviour. But what do the lawyers want from the neuroscientists? And are the scientists in a position to deliver the goods? In this chapter I will explore what the neurosciences can offer legal analysis and discuss the practical and theoretical limits we face in looking for the roots of criminal action. In particular, I will consider the extent to which we can ascribe intention to a person's action and ask whether neuroscience can offer anything legally relevant to this fundamental question.

How should we understand the intention behind a person's actions? It is a requirement of many criminal law prohibitions that a person should intend to cause the relevant harm; the requirement accords with the principle famously set down by Edward Coke: '*actus non facit reum nisi mens sit rea*', (the act is not guilty unless the mind is also guilty). Where, then, do we look for the intent? Recently legal scholars have begun to consider psychology and the neurosciences for guidance in interpreting the behaviour and motivation of a person who has committed a crime. However, we need to find the appropriate level at which to talk about psychological constructs that are of relevance to the law. Does it make sense to talk about the passage of individual action potentials along the axon of a neuron? Or should we talk more abstractly about higher-level goals? Dennis Patterson[1] has argued for the latter position, that we should talk in terms of *reasons* for action. I will examine this view and

* I am grateful to Bebhinn Donnelly-Lazarov for helpful comments.

[1] Dennis Patterson, 'Neuroscience and the Explanation of Human Action', this volume, Chapter 1.

present a counterargument based on what is currently known about the neuroscience of intention.

Mens Rea and *Actus Reus*

The law has from its earliest days had a dual approach to action. A defendant must have performed an act and must have had a culpable state of mind in doing so. If a person is behind the wheel of a car that injures a pedestrian, she has surely committed an injurious act – an *actus reus* – but for the driver to be criminally liable she must have intended the relevant act, or have had another legally recognised culpable state of mind, for example, by choosing to drive after drinking alcohol or by choosing to drive with excessive speed. This state of mind, the *mens rea*, must be demonstrated if the driver is to be found liable for the injury to the pedestrian. It is possible to have each of these two components separately: the driver's car may have mounted the pavement after being struck by another car (*actus reus* but no *mens rea*), or the driver may have left the house that day with the intention of running over her ex-partner but decided against it at the last moment (*mens rea* but not followed through with *actus reus*). Similarly, the offensive act, in appropriate circumstances, could be one of omission (failing to maintain the car's brakes) or one of possession. In any case, there must be a culpable state of mind, an intentional quality to the action, for liability to hold. I will not deal further with omissions here because the principles can apply equally to deliberately acting and to deliberately nonacting. I will consider how neuroscience might usefully shed light on the intentional states of criminal law.

For legal purposes, an 'act' has a fairly clear definition; it is the physical aspect of the crime; a person's bodily movements. These movements might involve a gross motor act such as a kick, a punch, or a tilt of a steering wheel, or there might be a movement of the vocal apparatus or the typing fingers, in the case of offences of language. In any case, an overt change of the person's bodily state must lead to the harm, even when the harm is separated in time or directness from the act.[2] This

[2] In *Thabo Meli and Others vs R* [1954] 1 WLR 228 (PC), the appellants were found to have killed a man, not through a blow to the head, which the victim survived, but from the attempt to dispose of (what they believed to be) the corpse by rolling it off a cliff. Rolling a dead body off a cliff does not constitute murder, but the chain of the day's events included acts intended to kill. So coincidence of cause and effect may be flexible.

definition protects cognitive liberty; the freedom to think whatever one pleases as long as no illegal act results.[3] Although there are some toothsome scholarly arguments around whether some acts truly count as acts,[4] it is generally clear when a person's body has caused an event to occur.

However, the body may not always be under the person's control, and involuntary movements, such as tics or spasms, and acts committed while asleep or unconscious do not generally occasion legal liability, unless the person knowingly neglected to prevent those movements (such as if a person neglected to take medication that would suppress spasms). What is missing, typically, in these cases is the required mental element, the intention, foreknowledge, suspicion that a given act would be likely to cause a given harm or bring about a particular state of affairs.

Let's take intention as one such state of mind. Where does that mental intention come from? Here I will examine the arguments that link mental causation to the physical properties of the person and her environment and will review the ways in which the new neurosciences help us to understand how to ground the metaphysical theories in the functions of the brain and body.

The Philosophical Status of Neuroscience

Most modern neuroscientists would consider themselves heirs to the tradition of *materialism*. That is to the idea that all phenomena in existence, including mental processes, arise solely from the properties of matter. A materialist would hold that there is no need to invoke a higher power or spirit to explain our inner life, and that a full understanding of mental processes will eventually come from understanding the properties of the matter that constitutes the person. This materialism stands in contrast to the *dualism* of philosophers such as Rene Descartes, who could not reconcile material and mental phenomena. A practical problem in dualism is in explaining how nonmaterial phenomena are able to influence the physical apparatus of the brain and thereby the body. Descartes chose the pineal gland as the site in the body where

[3] J.-C. Bublitz, 'My mind is mine?! Cognitive liberty as a legal concept'. In E. Hildt & A. Francke (Eds.), *Cognitive Enhancement* (Dordrecht: Springer, 2013).

[4] In a famous example, the aforementioned Edward Coke unsuccessfully defended the preacher Edmund Peacham, who was found in possession of a sermon that had never been, and apparently would never be, delivered: *The Case of Edmund Peacham* (1615) 2 How. St. Tr. 869.

nonmaterial mental processes interact with the physical body. But identifying a location in the brain is not the same as identifying a process, and there still do not appear to be any good theories in the dualist tradition of how material and nonmaterial phenomena interact.

Modern materialism is more properly called *physicalism*, the more precise name acknowledging that there is more to the physical realm than matter itself. The interactions among material phenomena also contribute to the functions of the system. So a computer processor can be understood in terms of the lattice of gates that is etched into it, but the true nature of the processor is only revealed when the gates are configured to run a program. A complete physicalist theory of mental processes may possibly draw on the electromagnetic properties of the cells that constitute the body, on the dynamics of thermal or metabolic transport around those cells, or on the relationship between the person and the surrounding environment. Although physicalism is somewhat difficult to define,[5] there is at least the general sense that the physical laws of the universe contain within them the basis for all physical and mental phenomena.

A well-celebrated conundrum in modern neuroscience is the status of *determinism*. Determinism is the philosophical position that the current state of any system (the brain, the plumbing, or the whole universe) is completely determined by its state immediately prior. That is, if we could know everything about the state of the brain in one instant, including the position and momentum of every particle, we could in principle know how the state of the brain will unfold over time and could therefore predict how the person will act at all times in the future.[6] Although there are of course practical and physical barriers to complete knowledge of a system like the brain, such as the famous uncertainty principle of quantum mechanics, true determinism excludes the idea of uncaused events. However, these uncaused events are precisely what free will requires because otherwise the future choices of the brain would be completely constrained by the developing state. The conundrum of determinism continues to trouble neuroscientists and philosophers alike. This chapter

[5] A. Ney, 'Defining physicalism' (2008) 3/5 *Philosophy Compass* 1033–1048.

[6] This is a gross oversimplification, of course. Rummens and Cuypers pick this apart, establishing that there are different forms of predictability and suggesting that human agency allows for a certain degree of control over some forms of determinism. See S. Rummens and S. Cuypers, 'Determinism and the Paradox of Predictability' (2010) 72 *Erkenn* 233–249.

is not concerned to contribute to this debate but to address an associated one: the role of neuroscience in understanding intention.

Physicalism and Physical Incompleteness

The physicalist approach to intention treats the physical realm as a *closed system*, in that all phenomena exist within the system and are sustained by the system. For this to be true, we must find an explanation for mental phenomena that arises within the system and does not require an external intervention. An argument against physicalist approaches of this sort is that there just does not seem to be a good way to go about the required method within our current knowledge of the physical universe. David Chalmers makes the point well, famously drawing the distinction between the 'easy problems' of the philosophy of mind, which are usually problems that can be put as questions of physiology or information processing, and the 'hard problems' of our sense of perception and cognition, for which we do not have a good way to frame a question in physical terms.[7]

Despite these apparent limitations, many legal scholars are now turning to the neurosciences to reduce the uncertainties in legal processes. And, despite the problems that neuroscience faces in defining the proper substrate for analysis, nevertheless there is a body of knowledge in the brain sciences and a set of methods for developing ideas, which may add significantly to the legal analysis of specific cases. Neuroscience may not answer the big questions about free will, for example, but it has value in understanding a person's intentions and motivations and in determining a proper outcome. Following are some applications.

Mitigation and Treatment

The nature and the severity of many crimes may indicate that a person should be committed for treatment for their own good and for the good of others. For example, the M'Naghten Rules of the 1840s allow courts to test the state of mind of a defendant at the time of the act, based on the following principle:

[7] D. Chalmers, 'Facing up to the problem of consciousness' (1995) 2 *Journal of Consciousness Studies* 200–219.

> [T]o establish a defence on the ground of insanity, it must be clearly proved that, at the time of the committing of the act, the party accused was labouring under such a defect of reason, from disease of the mind, as not to know the nature and quality of the act he was doing; or, if he did know it, that he did not know he was doing what was wrong.[8]

The field of biological psychology offers some insights into the range of factors that may bias an otherwise healthy person towards certain types of behaviour. It is known that the personality dimensions of Conscientiousness and Neuroticism are moderately heritable,[9] suggesting a degree of predetermination in our actions: a person who regularly experiences worry and anger may have inherited a predisposition for these emotions from their parents. Likewise, intelligence and socioeconomic status share both genetic and environmental determination, although these factors may be mitigated through education.

None of the factors listed here have probative value. However, a compassionate judge may incorporate evidence from medical and scientific experts to understand how predisposing circumstances may mitigate the wrongness of an act. For example:

> When a neurological test can demonstrate that a defendant's thought process does not operate as one a judge would normally encounter – whether that is due to diminished culpability or a brain injury – a judge could consider that information, counteracting any implicit biases she may be holding about the defendant's obvious characteristics.[10]

I will return later to the issue of bias, but the hope here is that some kind of measurement of a person's brain state or mind state might help in arriving at an appropriate punishment or treatment for a person convicted of an offence.

Intention

The factors described are in a sense post hoc, in that a person must have committed an act before we can worry about them. However modern neuroscience allows us to delve deeper into the intentional quality of

[8] *M'Naghten's case* (1843) 10 Cl & F 200.

[9] R. Riemann, A. Angerleitner and J. Strelau, 'Genetic and environmental influences on personality: A study of twins reared together using the self- and peer report NEO-FFI scales' (1997) 65 *Journal of Personality* 449–475.

[10] B. Donald and E. Bakies, 'A glimpse inside the brain's black box: Understanding the role of neuroscience in criminal sentencing' (2016) 85 *Fordham Law Review* 481–502 at 500.

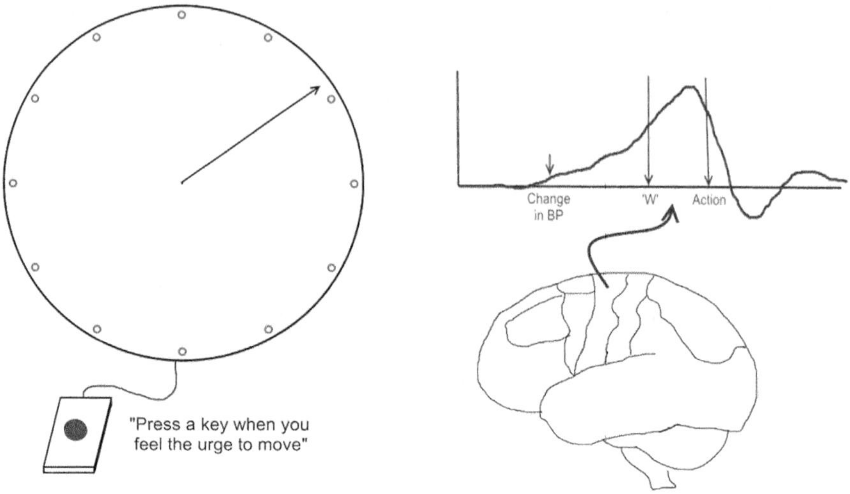

Figure 5.1 In the classic experiments of Benjamin Libet (e.g., Libet et al., 1983, at note 12), a participant is asked to press a key whenever he or she 'feels the urge (or the will) to move'. The participant is asked to note the time shown on the clock face at the time of the 'will' – this is known as the W-judgement. The Bereitschaftspotential (BP), which is a measure of the electrical activity of the brain around the time of movement, shows a deflection leading up to the time of an action (the button-press), at which point the potential returns to baseline. The W-judgement precedes the movement, as we would expect. However the key finding of Libet's work is that the BP begins its deflection some hundreds of milliseconds prior to the time when the participant thought he or she had decided to move. This is taken by some as evidence that 'free will' is no more than post hoc rationalising by an area of the brain for a decision already taken by another area.

action. Figure 5.1 shows a schematic version of a famous experiment in neuroscience, the so-called Libet Clock. Libet asked participants to press a key on a computer at a time of their choosing and to report what time was on the clock when the participant 'felt the will to move' (called the W-judgement).[11] The participants consistently reported that the time of the W-judgement occurred before the time of the actual key-press, which would, of course, be necessary, as the intention must precede the act. However, Libet's experiment revealed something more interesting: the activity of the brain showed a clear shift ahead of the W-judgement,

[11] B. Libet, C. A. Gleason, E. W. Wright and D. K. Pearl, 'Time of conscious intention to act in relation to onset of cerebral activity (readiness-potential) the unconscious initiation of a freely voluntary act' (1983) 106 *Brain* 623–642.

measurable up to half a second before the key-press. Because this shift in activity preceded the participants' experience of deciding to press the key, many scientists and philosophers have taken this as evidence that our sense of intention is no more than the conscious brain developing a post hoc explanation for an action that the nonconscious brain has already decided on. Variants of Libet's original experiment have confirmed the basic premise, that brain activity changes before the 'decision' to act, and sophisticated neuroimaging can detect activity up to ten seconds before a free choice that accurately predicts that choice.[12]

Seductive Allure

It must be admitted that there remains a certain attraction in pinning messy legal realities onto a seemingly clean scientific field such as neuroscience. In fact, the so-called seductive allure of the neurosciences may bias people into believing evidence without properly questioning it, which would undermine the use of such evidence in court.[13] Although there is some doubt about the solidity of this finding,[14] nevertheless there is something evocative about brain images that lends credibility to the facts of a case. This is not unique to the brain sciences, of course: many miscarriages of justice have occurred because courts have been poorly equipped to properly examine scientific evidence presented by expert witnesses.

However, the allure of the brain sciences, and the seeming solidity of the evidence and its presentation, induces a false confidence in people who handle such evidence. For example, the following passage was written by an American judge and a legal clerk (emphasis mine):

> Without disregarding the criminal justice system's ability to hold those accountable for their actions, neuroscience can be utilized to demonstrate that certain actions may actually be the result of developmental problems

[12] C. S. Soon, M. Brass, H. J. Heinze and J. D. Haynes, 'Unconscious determinants of free decisions in the human brain' (2008) 11 *Nature Neuroscience* 543–545.

[13] Classically this was reported by D. Weisberg, F. Keil, J. Goodstein, E. Rawson and J. Gray, 'The seductive allure of neuroscience explanations' (2008) 20 *Journal of Cognitive Neuroscience* 470–477, but see also D. P. McCabe and A. D. Castel, 'Seeing is believing: the effect of brain images on judgments of scientific reasoning' (2008) 107 *Cognition* 343–352, and D. Weisberg, J. Taylor and E. Hopkins, 'Deconstructing the seductive allure of neuroscience explanations' (2015) 10 *Decision Making* 429–44.

[14] Some methodological and other concerns were raised by M. Farah and C. Hook, 'The seductive allure of 'seductive allure' (2013) 8 *Perspectives on Psychological Science* 88–90.

> associated with the brain, like the effects of complex trauma on children. *A judge may also use neuroscience to combat her implicit biases, which have ways of manifesting themselves in the courtroom and therefore need to be explicitly acknowledged.* Neuroscience can offer additional insight into a defendant's thought process and accordingly provide a means for the judge to address and correct those biases.[15]

I would argue that reliance on neuroscientific evidence is itself a source of bias. No scientific fact is beyond questioning, and we should be as prepared to acknowledge and to challenge neuroscience evidence as we are to challenge evidence from bite marks or from DNA.[16] Given the newness of the relationship between the law and the neurosciences, there is a responsibility to get this right from the outset. Counsel must be given the confidence to present, and to challenge, evidence derived from the brain sciences, and expert witnesses must ensure that evidence presented in court is digestible without being overly simplified.

Practical Problems in Finding Neuroscientific Explanations

The previous section sounds rather optimistic. The neurosciences can tell us so much about predisposition and intention, so why do we not hand over judicial decisions to the scientists? There are several practical problems in reading brain states and in relating brain states to mind states.

Complexity and Redundancy in the Brain

There are currently significant gaps in our knowledge of how action plans arise in the brain and are thereby enacted. The neuroanatomy of the motor system is well known. The primary motor cortex, lying in a strip at the back end of the frontal cortex, projects to the spinal cord and from there to the muscles. Although there are some controversies about exactly how information is processed in the primary motor cortex,[17] there are helpful regularities such as a somatotopic mapping from brain areas to

[15] Donald and Bakies, note 10, at 482.

[16] For concerns around the use of bite marks, see G. V. Reesu and N. L. Brown, 'Inconsistency in opinions of forensic odontologists when considering bite mark evidence (2016) 266 *Forensic Science International* 263–270. There is a readable account of problems with DNA evidence in P. A. Smith, 'When DNA implicates the innocent' (2016) 314 *Scientific American* 11–12.

[17] N. J. Davis, S. P. Tomlinson and H. M. Morgan, 'The role of beta-frequency oscillations in motor control' (2012) 32 *Journal of Neuroscience* 403–404.

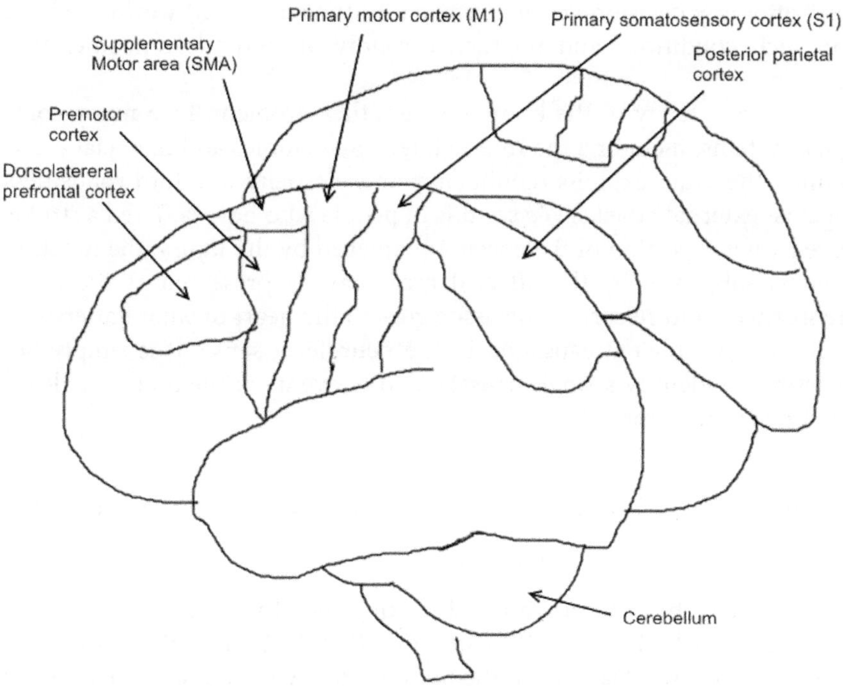

Figure 5.2 The key motor-related areas of the brain. The primary motor cortex (M1) is the principal outflow of information to the muscles. However, other areas are involved in planning actions (dorsolateral prefrontal areas and the posterior parietal cortex), and still others are involved in translating these plans into muscle dynamics (supplementary motor area, premotor cortex). The cerebellum and the primary somatosensory cortex (S1) monitor and guide ongoing action. The large amount of real estate devoted to action planning and coordination points to the complexity of information processing relating to action in the brain.

muscles, meaning that changes in activity measured in these areas have a predictable and predictive relationship with changes measured in the muscles. 'Upstream' of the primary motor cortex it becomes increasingly difficult to separate clear motor information from cognitive function. The higher motor areas of the premotor cortex (PMC) and the supplementary motor area (SMA) deal with higher-order plans related to goals and motivations, rather than to specific actions. The motor areas of the brain are shown in Figure 5.2. The experiments of Libet and Haynes and others, mentioned earlier, took place under rigidly controlled conditions and with detailed and accurate knowledge of the state of the brain before and after any decision was undertaken. In the real (legal) world, we have no such conditions and no such certainty around the outcomes that we see.

The complexity of the brain adds another problem. Like many complex systems, including the Internet, termite colonies and financial economies, the brain exploits redundancy and plasticity to afford robustness against external attacks. We see this in people who have suffered a stroke. Even when a section of the cortex is damaged by the injury, the function that is subserved by the affected area may be preserved as the brain reorganises and relearns. This issue goes to the heart of what Patterson[18] describes as efficient causation in the neurolegal sense. It is simply not possible to identify a single causal chain of events in the brain that leads to a given outcome.

Philosophical Problems in Finding Neuroscientific Explanations

Traditions in Neuroscience

So far I have been careful around making the obvious link between the mind and the brain. There is no question that the brain is the centre of all mental processes. Damage to the brain is clearly linked to altered mental processing, and even Descartes looked in the brain for the seat of his dualist interactions. However, it would not be accurate to suggest that the brain is the closed system that we are looking for. The most obvious objection to this view is that the brain does not exist on its own, but is connected via a variety of tissues to the sensory organs, to the muscles, and to the internal organs of the body. In a trivial sense, the brain is crucially situated in the body and is immersed in the sensory milieu. Notwithstanding philosophical conundrums such as the brain-in-a-vat thought experiment of Gilbert Harman,[19] the brain is part of the body and is indivisible from it.

A branch of behavioural psychology suggests that not only should we consider the brain in the context of the whole body, but also further that the functions of the brain and body only make sense in relation to the flow of information passing around and through them. This field of

[18] Patterson, n.1.

[19] This was a neat thought experiment devised by Harman in G. Harman, *Thought* (Princeton, NJ: Princeton University Press, 1973) and was developed by Hilary Putnam in H. Putnam, *Reason, Truth and History* (Cambridge: Cambridge University Press, 1981).

ecological psychology observes that all of the information needed to behave freely in the world is available directly from the sensory arrays, and that little, if any, information needs to be stored in a memory representation.[20] Extending this further, the programme of *embodied cognition* seeks to understand mental function in relation to the environment and to available behaviours.[21] For example, mentally rotating an image of a hand is faster than mentally rotating an image of blocks, which is consistent with our familiarity with our own hands.[22]

The difference between many fields of cognitive science comes down to what quality of the universe is being minimised or maximised. Many researchers who study the motor system believe that movement errors should be minimised,[23] while those who study the sensory systems believe that some measure of sensory fidelity should be maximised.[24] The ecological branch focuses on the amount of information processing that is needed, believing that it is wasteful for the brain to simulate or calculate information that is cheaply available in the environment.[25]

The Disunity of Neuroscience

All this is to say that the proper substance of neuroscientific analysis has not been sufficiently established. Researchers in different fields, or in different traditions, look at the cognitive architecture of the brain and see different things. Many other researchers are less concerned with the information processing within the brain, as they are with the tools they have for extracting that information. People who study language functions almost invariably use electroencephalography (EEG), while those who study brain-wide cognitive of memory processes may reach for functional magnetic resonance imaging (fMRI). The wholeness of our

[20] J. J. Gibson, *The Ecological Approach to Visual Perception* (Boston, MA: Houghton Mifflin, 1979).

[21] E. Thompson and F. Varela, 'Radical embodiment: neural dynamics and consciousness' (2001) 5 *Trends in Cognitive Sciences* 418–425.

[22] S. Kosslyn, G. DiGirolamo, W. Thompson and N. Alpert, 'Mental rotation of objects versus hands: neural mechanisms revealed by positron emission tomography' (1998) 35 *Psychophysiology* 151–161.

[23] C. Harris and D. Wolpert, 'Signal-dependent noise determines motor planning' (1998) 394 *Nature* 780–784.

[24] D. C. Knill and A. Pouget, 'The Bayesian brain: the role of uncertainty in neural coding and computation' (2004) 27 *Trends in Neurosciences* 712–719.

[25] D. N. Lee and P. Reddish, 'Plummeting gannets: a paradigm of ecological optics' (1981) 293 *Nature* 293–294.

understanding of the brain and its activity is threatened by the disparity between the methods and the theories used to understand it.

Can there be unity in the fields of behaviour and neuroscience? I believe there can. The philosophical bedrock comes from the notion that we live in a *computable* universe where systems progress according to lawful rules.[26] Those rules come from the laws of physics, and particularly the laws that apply at the scale of our body size, and to the substances that compose our bodies. For example, Fick's laws describe the diffusion of chemicals in a fluid, Maxwell's equations relate different forms of electromagnetic energy, and Boltzmann (following Newton) developed a basis for understanding entropy. With these laws in place, and a sense of their dynamics, we have the basis for understanding brain processes in relation to each other, which is the beginnings of a theory. However, there is some considerable work to be done to get to that point.

Possible Solutions

The Need for Better Neuroscience

Where does this epistemic obstacle course leave us? Many of the issues raised are practical ones, and the solution may lie in better technology and better theories. Pushing the envelope of brain imaging allows us to see intentions develop, several seconds before action.[27] Another solution may be to attack the problem from another angle. The same advances in neuroimaging that showed us the anatomy of intention may also be used to probe the content of memory. The field of lie detection is relatively well established, but is held back by the legal and moral questions around extracting the true state of another person's mind. The right to cognitive integrity is thought by many to be just as strong a human right as the right to bodily integrity, the violation of which is treated with the utmost seriousness in most jurisdictions.

Abstraction and Emergence

We may also learn from the fields of science that deal daily with the most complex systems. The science of the climate is just such a field. The large-scale weather systems that gyre across the oceans are the result of

[26] D. Deutsch, *The Fabric of Reality* (London: Penguin Books, 1997).
[27] Soon et al., note 12.

uncountably many uncountably tiny interactions among molecules of water and gas and the rays of the sun. Yet we can ask simple, direct questions of them, such as: 'Should I wear a coat on Sunday?' or: 'Will there be snow at Christmas?' The weather scientists add all available information into their computers and deliver their verdict: 'unlikely to be necessary', or: '80 percent probable'. People who ask and answer questions at the level of clothing decisions do not need to know the physical properties of water, or the likelihood of every possible interaction between every molecule of water and every possible solar ray that passes through the atmosphere. Instead we are comfortable with the statistical nature of the physical processes, and with the probabilistic nature of the predictions that arise from them. Similarly, we do not need to know the exact pathway of every neuron in a person's brain, nor the distribution of neurotransmitters and peptides, to know whether our friend is sad to have ended her relationship with her girlfriend. Again, we are comfortable with approximations and probabilities. In both of these cases we develop an appropriate level of abstraction that suits the form of the question we want to ask. This does not deny the 'truth' of the most detailed level of analysis, but serves to remind us that most human cognition operates at a level of abstraction that might best be termed 'beyond reasonable doubt'.

The preceding touches on what has become known as *emergence*, or the notion that the proper level of analysis is not the tiniest level of physical detail, but at a higher level, where matter and forces interact lawfully to produce larger-scale effects. As John Blodwell puts it:

> There is an underlying, unspoken assumption about what has come to count for us as existential bedrock, namely that at bottom 'reality' is stuff in space, evolving and interacting in accordance with the laws of physics ... Ontological status then favours objects rather than operations.[28]

However, Blodwell argues that mental phenomena are not properties of the constituents of the brain (neurons, glia, cerebrospinal fluid, etc.) but of the interactions among those parts. These phenomena emerge when the parts communicate with each other.

Crucial to the discussion of emergence is the sense that the emergent phenomena are stable in some way. We are happy to believe that the

[28] J. Blodwell, 'Physicalism and emergence' (2011) 18 *Journal of Consciousness Studies* 6–31 at 7.

'person' (collection of cells and fluids collected within a matrix of skin and clothes) who stands in court bears some relation to the 'person' (collection of cells, etc.) that existed at the time of a supposed crime. And further that the same collection of cells will emerge from a period of corrective detention, chastened but improved. This temporal coherence gives meaning to dynamic systems such as personhood or cognitive processing.

Reasons

Patterson has argued that the preceding examples focus on an inappropriate level of detail. Instead, the philosophical constructs of *reasons* are the best level on which to consider issues of motivation. What are reasons (in this context)? Robert Audi[29] defines a range of different types of reasons, which cover the different senses in which people mean the term. However the classes of reasons that concern us here are *normative reasons* and *explanatory reasons*. Normative, or justifying, reasons are reasons that make it sensible or likely that a person will act in a certain way. So the reason why I check if my dinner guests may be vegetarian is normative: I would not want to offend someone by presenting her with a steak if she does not eat meat. Actions result from a weighing up of the positive and negative normative reasons for those actions. By contrast, explanatory reasons are the reasons that led the person to commit the act. These are the causes (the Aristotelean *efficient causes* as Patterson notes) of the action by that specific person on that specific day.

Audi later[30] develops and elaborates the notion of reasons, relating them to different conceptions of moral virtue, or understanding the chains of reasoning (or *schemata*) that derive from basic reasons. However, a limitation of the use of reasons in considering human behaviour is that they are opaque to external observers. The person who possesses the reason may have rational and well-justified reasons for her behaviour, but we still need that person to tell us what they are. So normative reasons may be guessed at from the time or the culture in which a person lives, but the explanatory reasons for a specific act may not be obvious unless a person chooses to reveal them.

Reasons are not, therefore, a directly helpful way to judge *mens rea*. Could they have any other uses? Examination of a person's reasons for

[29] R. Audi, 'Acting for reasons' (1986) 95 *The Philosophical Review* 511–546.
[30] R. Audi, *Reasons, Rights, and Values* (Cambridge: Cambridge University Press, 2015).

action may reveal that a person believed they had acted on behalf of God, or in the cause of preserving the racial 'purity' of the nation. Neither of these is a good reason (or rather, an exculpatory reason), but the court may take these beliefs (or delusions) into account when sentencing for a crime.

Conclusions

Dennis Patterson[31] argues that the appropriate level of analysis for criminal neuroscience is that of reasons. I agree with Patterson that current neuroscientific explanations of action are unsatisfying, but I would further suggest that we keep trying. Steering away from physicalist explanations does not solve the problem the legal system has wrestled with for centuries: whether a person had the relevant guilty intention (*mens rea*) at the time when they committed the guilty act (*actus reus*).

Law and neuroscience are natural friends. Legal scholars will find it profitable to understand the basis of voluntary action and its underpinnings in the brain. However, the lawyers must also understand the limits of current understanding about the functions of the brain, both in terms of the technological limits of the machines we use to probe the brain, but also in terms of the questions we are able to ask of the brain. Conversely the scientists will benefit greatly from contact with a field in which their knowledge can be properly applied because new relationships breed new questions, and constant questioning is what drives science to improve. For their part, scientists must remain open to these enquiries and must provide answers that engage with the question without skipping detail and without implying certainty or consensus. The so-called seductive allure of brain-based explanations may introduce bias to legal proceedings without adding to the scientific value, and to do so devalues the law and the neurosciences.

At the start of this chapter I asked what it was that lawyers want from the neurosciences, and whether the scientists were able to provide what is asked of them. The question: 'Did this person intend to commit this act?' is a vital one for criminal law. Neuroscience is not equipped to answer this question yet. The roots of intentional behaviour undoubtedly lie in the brain, but extracting that signal from the vast architecture of the

31 Patterson, note 1.

brain is certainly beyond current technology. Asking people to 'intend' in a lab, as in the paradigms of Benjamin Libet, or of Soon et al., reveals a network of brain areas that work together at different stages of the process. However, none of these approaches can answer the more legally relevant post hoc question: for the act that was committed by this person, did the person intend or not intend the prohibited outcome? Now the question becomes one of truth-detection, and again the neurosciences do not have a good, reliable means of separating truthful from deceitful responses. So, from these methods, the scientists cannot do what the lawyers seem to want of them: to answer the question of intention. Neuroscience undoubtedly has much to offer the legal system. In this specific realm, science can, as yet, offer very little to aid the court, but things may change.

PART II

Epistemic Disputes

What Does Neuroscience Tell Law about Responsibility?

6

Lying, Deception, and fMRI: A Critical Update

MICHAEL S. PARDO*

Among the topics at the intersection of law and neuroscience – or "neurolaw" – the use of fMRI as a lie detector has been one of the most prominent for scientists, academic commentators, and popular media. In one sense, this is understandable. Determining whether someone is speaking sincerely or is lying is a tremendously important legal topic, and it is one that is notoriously difficult to ascertain in the context of litigation. The ability to bypass the usual complex inferential process of assessing testimony with "direct" evidence of whether someone is lying is thus understandably attractive.[1] Moreover, fMRI lie detection does not appear to raise the same sorts of thorny philosophical and empirical issues that affect other proposed uses of neuroscience in law – such as those involving free will,[2] ascriptions of criminal responsibility,[3] or even

* My thanks to Bebhinn Donnelly-Lazarov and Dennis Patterson for helpful comments and to Dean Mark Brandon and the Alabama Law School Foundation for generous research support.

1 The idea that fMRI offers "direct" access to lies in the brain (an idea advanced by some scholars and proponents of the technique) is based on a mistaken conceptual assumption about the evidence. As with more traditional lie-detection techniques such as polygraph, fMRI lie detection also relies on supposed correlations between lying and physiological changes in the body. Physiological changes in the brain no more constitute "lies" than do an increased heart rate or sweating. For a discussion of this conceptual issue in the context of fMRI lie detection, see Michael S. Pardo and Dennis Patterson, *Minds, Brains, and Law: The Conceptual Foundations of Law and Neuroscience* (New York: Oxford University Press, 2013), pp. 105–06. Moreover, even when a witness's sincerity is certain, this does not eliminate the inferential work necessary to evaluate the witness's testimony. Sincerity is not the same as veracity – it is also necessary to evaluate whether the witness is sincere but mistaken.

2 Adam J. Kolber, "Free Will as a Matter of Law," in Dennis Patterson and Michael S. Pardo (eds.), *Philosophical Foundations of Law and Neuroscience* (Oxford: Oxford University Press, 2016), p. 9.

3 Stephen J. Morse, "The Inevitable Mind in the Age of Neuroscience," in Dennis Patterson and Michael S. Pardo (eds.), *Philosophical Foundations of Law and Neuroscience* (Oxford: Oxford University Press, 2016), p. 29.

the related issue of mind reading.[4] Rather, on the surface at least, fMRI lie detection appears to involve a straightforward application of probative evidence being used to prove a relevant and important factual issue in law.[5] Thus, the understandable attention devoted to the topic.[6]

In *Minds, Brains, and Law*, Dennis Patterson and I examined the scholarly literature and judicial opinions on the topic.[7] We discussed several empirical and conceptual issues that affect the use of such evidence in law. Although the empirical limitations with the evidence have been explored in detail, the conceptual issues with the evidence have received considerably less attention. Consistent with the general theme of that book,[8] we focused in detail on the conceptual issues and the challenges that they raise for law.

In this chapter, I will focus on two conceptual issues and examine several studies that have been published since the publication of our book. The conceptual issues concern: (1) the distinction between deception and lying and (2) the concept of lying itself (or the criteria for what constitutes a lie). As with the array of studies that we examined in our book, the more-recent studies also face serious limitations because of these conceptual issues. Part I summarizes the empirical and conceptual issues concerning fMRI lie detection. Part II illustrates these issues by discussing a concrete example: the attempted use of such evidence in the case of *United States v. Semrau*.[9]

[4] Gideon Yaffe, "Mind-Reading by Brain-Reading and Criminal Responsibility," in Dennis Patterson and Michael S. Pardo (eds.), *Philosophical Foundations of Law and Neuroscience* (Oxford: Oxford University Press, 2016), p. 137.

[5] Despite this appearance, several difficult conceptual issues underlie the application of this evidence in law. Moreover, the fact that the conceptual issues are typically presupposed (and harder to detect) leads to some of the confusion about them. For discussion, see Pardo and Patterson.

[6] In another sense, however, the focus on lie detection is somewhat strange given the historical reluctance courts have shown toward lie-detection evidence in general. This reluctance, moreover, often concerns issues beyond perceptions about the reliability of such evidence. For excellent discussions, see D. Michael Risinger, "Navigating Expert Reliability: Are Criminal Standards of Certainty Being Left on the Dock?" (2000) *Albany Law Review* 99 at 129–30; Frederick Schauer, "Lie-Detection, Neuroscience, and the Law of Evidence," in Dennis Patterson and Michael S. Pardo (eds.), *Philosophical Foundations of Law and Neuroscience* (Oxford: Oxford University Press, 2016), p. 85.

[7] Pardo and Patterson, pp. 79–120.

[8] Ibid. at p. xxi. ("The conceptual issues focus on the scope and contours of the concepts being employed in claims involving law and neuroscience.")

[9] *United States v. Semrau*, 693 F.3d 510 (6th Cir. 2012); *United States v. Semrau*, 2010 WL 6845092 (W.D. Tenn., June 1, 2010).

Finally, Part III summarizes recent studies and explains the conceptual problems at the heart of the experiments.

Functional MRI Lie Detection: Empirical and Conceptual Issues

The basic idea behind fMRI lie detection is that identifiable areas or patterns in the brain will correlate more with lying than with truthful behavior. Based on the strength of these correlations, so it is claimed, reliable inferences may be drawn about whether a person is lying or telling the truth on a particular occasion (or with regard to a particular subject matter).

The research supporting such claims depends on experiments such as the following.[10] Subjects are placed in an fMRI scanner, and they are presented (for example, on a computer screen or through headphones) with a series of binary questions (for example, yes–no, true–false, heads–tails, and so on). Subjects are instructed to answer by pressing one of two buttons, each corresponding to one of the possible answers. Subjects are also instructed either to "lie" or to answer truthfully about an array of different topics such as biographical information, daily events, numbers, playing cards, past events, or under which object money was hidden. While subjects answer, the scanner measures brain activity indirectly by measuring blood flow to areas of the brain.[11] Scanning data is then processed and translated into a brain "image," which is a visual representation of the underlying statistical measurements of blood flow.[12]

Based on these studies, researchers reported several positive findings on fMRI lie detection. Initial studies examined group results to explore

[10] For examples of the several studies employing this paradigm, see Pardo and Patterson, pp. 82–85.

[11] The technique depends on a number of important scientific principles – most important, the BOLD ("blood oxygen dependent level") signal to infer conclusions about brain activity. The relationship between the signal and brain activity depends on two ideas: when hemoglobin in blood delivers oxygen to the brain it becomes "paramagnetic" and will disrupt a magnetic field (such as one created by an MRI scanner), and when activity increases in a particular brain region, blood flow will increase in order to supply more oxygen. For an excellent overview, see Adina L. Roskies, "Brain Imaging Techniques," in Stephen J. Morse and Adina L. Roskies (eds.), *A Primer on Criminal Law and Neuroscience* (New York: Oxford University Press, 2013), pp. 37–71.

[12] Ibid.; Teneille Brown and Emily Murphy, "Through a Scanner Darkly: Functional Neuroimaging as Evidence of a Criminal Defendant's Past Mental States" (2012) 62 *Stanford Law Review* 1119.

whether activity in any brain areas in general correlated with lying.[13] These studies identified several brain areas, including the prefrontal and the anterior cingulate cortices.[14] Subsequent studies then focused on individual-level determinations, sometimes reporting impressive results. For example, one study described a 90 percent success rate in identifying which object was stolen in a mock-crime experiment.[15] A second study reported a 100 percent success rate in identifying the participants in a mock crime (9 out of 9), along with a 33 percent rate in identifying "no crime" participants (5 out of 15).[16]

Several empirical issues, however, limit the utility of such research for providing a foundation for evidence in legal settings. These issues include (1) the small sample sizes of the subjects tested, (2) the lack of diversity among those tested, (3) the time differences between experimental and litigation settings, (4) differences among types of lies, (5) the low-stakes conditions for experimental subjects, (6) the fact that the identified brain regions are also correlated with behaviors and activities other than lying, (7) the effectiveness of countermeasures, and (8) the procedures used for constructing brain "images."[17] The fundamental empirical question in a litigation setting is how likely a particular answer given by an expert reporting the results from fMRI technology is likely to be accurate when applied to a particular person, answer, or subject matter. Each of the issues listed earlier affect the answer to this question and thus the inferences that may be drawn from such evidence.

In addition to the empirical issues, a number of *conceptual* issues also affect the quality of fMRI evidence in this context. Although the conceptual issues are less noticeable and have received considerably less attention in the literature, they are both significant and problematic. The array of empirical issues noted earlier takes place within a conceptual

[13] Pardo and Patterson, pp. 85–86. [14] Ibid. at 85.

[15] F. Andrew Kozel et al., "Detecting Deception Using Functional Magnetic Resonance Imagining" (2005) 58 *Biological Psychiatry* 605.

[16] F. Andrew Kozel et al., "Functional MRI Detection of Deception After Committing a Mock Sabotage Crime" (2009) 54 *Journal of Forensic Science* 220.

[17] See Pardo and Patterson, pp. 97–99 (discussing these issues). Three recent articles provide excellent overviews of the empirical limitations regarding such evidence. Martha J. Farah et al., "Functional MRI-Based Lie Detection: Scientific and Societal Challenges" (2014) 15 *Nature Reviews Neuroscience* 123; Joshua W. Buckholtz and David L. Faigman, "Promises, Promises for Neuroscience and Law" (2014) 24 *Current Biology* R861; Henry T. (Hank) Greely, "Mind Reading, Neuroscience, and the Law," in Stephen J. Morse and Adina L. Roskies (eds.), *A Primer on Criminal Law and Neuroscience* (New York: Oxford University Press, 2013), p. 120.

framework that includes a number of presuppositions about what exactly is being measured in the experiments.[18] These issues are conceptual rather than empirical because they do not concern the empirical relationship between brain activity and behaviors; rather, they concern the concepts that are used to identity and pick out the target behaviors in the first place. For this reason, the conceptual issues are in an important sense prior to, or foundational for, the empirical issues. Confusing or erroneous conceptual presuppositions will render empirical claims built on top of them problematic. Experiments looking to confirm the brain activity that correlates with any variable *X*, in other words, need correct conceptual assumptions about what exactly constitutes *X*. In the lie-detection context, these conceptual issues include, for example, what constitutes lying and related issues.

In our book, we discussed several conceptual issues underlying fMRI lie detection. In this chapter, I will focus on two of them: the distinction between lying and deception, and the definition of lying employed or presupposed in the studies.[19]

The first conceptual issue concerns the distinction between lying and deception. In order to discover the brain correlates of lying and truthful behavior, the studies focus in particular on deception or an "intent to deceive" on the part of subjects.[20] In some cases, "lies" are defined in terms of "an intent to deceive." In one sense, this focus on deception makes sense because there is likely to be substantial overlap between lying and intending to deceive, and evidence that someone is acting deceptively makes it less likely that they are speaking sincerely. But the equation of lying with deception is also conceptually problematic because the two are not identical.[21] An intent to deceive is neither necessary nor sufficient to lie. It is possible to deceive without lying, and it is possible to lie without deceiving or even having an intent to deceive. An example of

18 Gabriel Abend, "What Are Neural Correlates Neural Correlates of?" (2016) *BioSocieties* 1–24.

19 Other conceptual issues with brain-based lie detection are discussed in Pardo and Patterson, pp. 99–115.

20 For examples, see Pardo and Patterson, p. 108.

21 Ibid.; Don Fallis, "Lying as a Violation of Grice's First Maxim of Quality" (2012) 66 *Dialectica* 563 ("several philosophers . . . have pointed out that there are lies that are not intended to deceive"), p. 563; Seana Valentine Shiffrin, *Speech Matters: On Lying, Morality, and the Law* (Princeton, NJ: Princeton University Press, 2014), pp. 19–26 (distinguishing lying and deception). Shiffrin also observes that "[t]he failure to distinguish carefully between lies and deception represents a defect in the empirical literature about lying." Ibid. p. 20, note 28.

the latter includes a witness who has been threatened into giving false alibi testimony for a witness.[22] The witness may give such testimony not intending to deceive but hoping the audience (the police, a judge, or a jury) will see through the false statements and refuse to rely on them. The witness will have undoubtedly lied, and the testimony will also fit legal definitions for perjury and false-statement crimes.[23] Importantly, however, the witness will not have had an intent to deceive. Thus, fMRI technology that *could* reliably measure deceptive intentions would nevertheless be *under*inclusive in measuring whether someone is lying or speaking sincerely.

This underinclusiveness is perhaps a relatively minor point, suggesting that the proposed technology may fail to pick up one category of lying behavior. Much more important, however, is the second conceptual issue – which reveals that the studies are significantly and problematically *over*inclusive in their conceptions of lying. This conceptual issue concerns what constitutes a lie and whether the experiments are measuring lies.

Although prototypical cases of lying may by easy to identify, the concept of lying has been notoriously difficult to articulate in terms of necessary and sufficient conditions. A great deal of recent philosophical analysis, however, has helped to clarify its scope and contours.[24] Most important, lying involves more than simply stating something that is

[22] Lies without an intent to deceive include the category of so-called bald-faced lies in which "a speaker 'goes on record' with something even though everybody knows it is false." Adam J. Arico and Don Fallis, "Lies, Damned Lies, and Statistics: An Empirical Investigation of the Concept of Lying" (2013) 26 *Philosophical Psychology* 790. See also Roy Sorensen, "*Bald-Faced Lies*! Lying Without the Intent to Deceive" (2007) 88 *Pacific Philosophical Quarterly* 251; Thomas L. Carson, "The Definition of Lying" (2006) 40 *Nous* 284. An example (from Carson) is a student whom everyone knows cheated and who brags openly about doing so. Suppose the Dean will punish the student only if he confesses (and the student knows this). Although everyone (including the Dean) knows the student cheated (and the student knows the Dean knows), the student denies cheating when questioned by the Dean. This is an example of lying, but the student is not trying to deceive anyone.

[23] For example, a witness commits perjury under federal law in the United States if, while testify under oath, he "willfully" asserts something "he does not believe to be true." 18 U.S.C. § 1621(1). "Willfully" in this context means making the statement with the intent to give false testimony. *Unites States v. Dunnigan*, 507 U.S. 87, 94 (1993). Similarly, a potential witness commits a federal crime in the United States when he "knowingly" and "willingly" makes a "false statement" to a government agent on a matter within the jurisdiction of the United States. 18 U.S.C. § 1001(a). Either of these crimes may be committed without having an intent to deceive.

[24] See, for example, the sources cited in note 22.

false or that a speaker believes to be false. There is also a social, normative component to lying that distinguishes it from other activities such as acting in a play, role-playing, playing a game, telling a joke, or using sarcasm, in which actors may say things they believe to be false. In our book, we relied in particular on recent work by philosopher Don Fallis, who has emphasized the "normative component of assertion that is necessary for lying."[25] As Fallis explains, the difference that makes saying "I am the Prince of Denmark" a lie when told at a dinner party but not when on stage during a play are the distinct norms of conversation that are in effect in the former. In the former, but not the latter, there is an operative norm against communicating something believed to be false. Prototypical cases of lying involve communicating something you believe to be false *while also* violating (or believing yourself to be violating) a norm of communication against doing so.[26]

This normative component is critically important in legal settings. Lying in such settings not only violates an operative communicative norm – it is also illegal.[27] This normative component, however, appears to be absent in the fMRI experiments. Subjects who give false answers in the scanner are not violating an operative norm against communicating something believed to be false; indeed, they were instructed to do just that.[28] Thus, the subjects do not appear to be lying. Therefore, the brain activity measured in such experiments does not appear to be brain activity accompanying lying.[29] Rather, it appears to resemble other types of nonlying activities noted earlier such as acting, joking, or playing a game. This is a conceptual problem that renders the empirical results

25 Don Fallis, "What Is Lying?" (2009) 106 *Journal of Philosophy* 29, 35.

26 Ibid. ("You believe that you make this statement in a context where the following norm of conversation is in effect: do not make statements that you believe to be false"). In a more-recent article, Fallis explained that this definition of lying may not cover some cases of lying by children or others who may be unaware of the norm against saying false things. See Fallis, 2012.

27 See note 22.

28 Note that this is distinct from whether the statements are "instructed lies." Someone can be instructed to lie and still lie – for example, an employee instructed by an employer to lie in a deposition. In doing so, the employee is still violating a norm against communicating false beliefs. When this norm is not present (such as in the experiments), then the instructed false statements are not examples of lying.

29 Among the fMRI literature, we mentioned as a notable exception one study whose experimental design did appear to create the possibility for lying by participants. Joshua D. Greene and Joseph M. Paxton, "Patterns of Neural Activity Associated with Honest and Dishonest Moral Decisions" (2009) 106 *Proceedings of the National Academy of Sciences* 12506; Pardo and Patterson, pp. 110–11.

from fMRI studies – and the inferences that may be drawn in legal settings – problematic. And it does so for reasons that are distinct from the litany of empirical issues noted earlier.

The next part illustrates the practical, real-world problems generated by this conceptual issue.

United States v. Semrau: An Application

The well-known and much-discussed case of *United States v. Semrau* provides a concrete example for illustrating the significance of the conceptual issues at the heart of fMRI lie detection.[30] The case is noteworthy because it is the first detailed attempt to introduce the results of fMRI lie detection at trial in a federal criminal case in the United States.[31] The case is also noteworthy because of the detailed and careful analysis provided by the trial[32] and appellate[33] courts in assessing expert testimony on the evidence, and the underlying scientific studies. The trial court excluded the evidence and the appellate court upheld the ruling, after each court discussed several empirical issues related to the evidence. Although the courts' reasons were sufficient to exclude the evidence, the conceptual issues (not discussed at either level) provide additional, foundational problems with such evidence. The conceptual issues are important to highlight because, in excluding the evidence, the courts in *Semrau* each acknowledged the possible admissibility of such evidence in the future. This part first discusses the details of the evidence and the courts' analyses and then discusses conceptual problems with the foundational studies underlying the evidence.

The defendant, Lorne Semrau, was charged with multiple counts of healthcare fraud and money laundering.[34] Semrau, a licensed psychologist, owned two companies that provided mental-health services to patients in nursing homes. The allegations against Semrau concerned a scheme to defraud Medicare, Medicaid, and other government programs

30 *United States v. Semrau*, 693 F.3d 510 (6th Cir. 2012); *United States v. Semrau*, 2010 WL 6845092 (W.D. Tenn., June 1, 2010).

31 In two other cases at the state-court level, parties have attempted unsuccessfully to introduce fMRI lie-detection evidence. *Wilson v. Corestaff Services L.P.*, 900 N.Y.S.2d 639 (May 14, 2010); *Maryland v. Smith*, Case No. 106589C (Maryland Cir. Ct. 201).

32 The district court adopted a detailed report and recommendation by Magistrate Judge Tu M. Pham. *United States v. Semrau*, 2010 WL 6845092 (W.D. Tenn., June 1, 2010).

33 *United States v. Semrau*, 693 F.3d 510 (6th Cir. 2012).

34 *Semrau*, 2010 WL 6845092, slip op. at pp. 1–2.

by submitting false and fraudulent claims for payments, totaling approximately three million dollars. In order to convict Semrau at trial, the prosecution had to prove that Semrau (1) "knowingly devised a scheme . . . to defraud," (2) "executed or attempted to execute this scheme," and (3) "acted with intent to defraud."[35]

In his defense, Semrau claimed that he did not intend to defraud with his billing practices, and that the mistakes in his billing resulted from confusion caused by unclear billing codes and instructions from insurance providers. To support his defense, Semrau sought to introduce fMRI results purporting to indicate that he was not lying when he claimed the billing practices at issue resulted from confusion rather than an intent to defraud.

Semrau attempted to introduce the evidence through the expert testimony of Dr. Steven Laken, the CEO of the company (Cephos) that produced the fMRI lie-detection results for Semrau.[36] According to Laken, Semrau was scanned according to a protocol developed by Cephos. The testing consisted of scanning Semrau while asking him three types of questions: neutral, control, and "specific incident questions" (SIQs). The neutral questions (e.g., "Are you over age 18?") were used to establish a baseline for Semrau; the control questions (e.g., "Do you gossip?" were used to fill empty space and were not considered part of the analysis; the SIQs concerned the allegations regarding the fraudulent billing practices. In the scanner, Semrau was instructed to "lie" or "tell the truth" when presented with SIQs and to tell the truth when presented with control and neutral questions.[37] According to Laken, the fMRI results suggest that Semrau was being truthful in his claims about the billing practices.[38]

The trial court concluded that Laken's testimony and the fMRI evidence should be excluded under Federal Rule of Evidence 702.[39] The court analyzed the evidence according to the factors articulated in

[35] Ibid. [36] Ibid.

[37] Ibid. at 6 ("In each fMRI scan, Dr. Semrau was visually instructed to 'Lie' or to tell the 'Truth' in response to each SIQ. He was told to respond truthfully to the neutral and control questions.")

[38] Ibid. at 6–8. In fact, Laken initially concluded that Semrau was being deceptive with regard to one set of question. He retested Semrau after concluding that these results may have been caused by fatigue. After retesting, he concluded that Semrau was being truthful. Laken also explained that his conclusion applies only to the general subject matter of the questions and not to any individual questions.

[39] Ibid. at 10–14.

Daubert: whether the evidence was (1) tested, (2) subject to peer review and publication, (3) had known error rates and standards of control, and (4) was based on principles and methods generally accepted in the relevant scientific community.[40] The court noted, with regard to the first two factors, that the fMRI technique had been tested in a number of studies (including some coauthored by Laken) that were published in peer-reviewed scientific journals.

The court, however, discussed several empirical problems with regard to the third factor: error rates and standards of control.[41] First, that court noted the error rates in the laboratory studies may not translate to real-world settings, for which there are no known error rates. Second, the court explained that there are no industry standards for such testing and that Laken appeared to deviate from his own protocol in testing Semrau. Third, Semrau differed in important respects from participants in the studies. Fourth, the studies themselves were limited in several respects, including (1) a small number of nondiverse subjects; (2) inconsistent results; and (3) the unknown effects of countermeasures. In addition, with regard to the fourth factor, the court concluded that the principles and methods of fMRI lie detection were not generally accepted among neuroscientists. For all of these reasons, the court held that the fMRI results were not sufficiently reliable to be admitted at Semrau's criminal trial.

Although the court excluded the evidence, the court explained that similar evidence may be admissible in the future, even if there remain unknown error rates for real-world settings:

> Should fMRI-based lie detection undergo further testing, development, and peer review, improve upon standards controlling the technique's operation, and gain acceptance by the scientific community for use in the real world, this methodology may be found to be admissible even if the error rate is not able to be quantified in a real world setting.[42]

Semrau was convicted at trial of three courts of healthcare fraud and appealed his conviction.[43] One issue that he raised on appeal was the exclusion of his fMRI evidence. The appellate court upheld the court's ruling:

> After carefully reviewing the scientific and factual evidence, we conclude that the district court did not abuse its discretion in excluding the

[40] *Daubert v. Merrell Dow Pharm., Inc.* 509 U.S. 579 (1993). [41] Semrau, slip op. at 13.
[42] Ibid. at 12, note 18. [43] *United States v. Semrau*, 693 F.3d 510 (6th Cir. 2012).

> evidence under Federal Rule of Evidence 702 because the technology had not been fully examined in "real world" settings and the testing administered to Dr. Semrau was not consistent with tests done in research studies.[44]

The court also explained that Federal Rule of Evidence 403 provided an independent basis for excluding the evidence because

> the prosecution did not know about the test before it was conducted, constitutional concerns caution against admitting lie detection tests to bolster witness credibility, and the test results do not purport to indicate whether Dr. Semrau was truthful about any single statement.[45]

Similar to the district court, the appellate court also noted the possible admissibility of such evidence in the future.[46]

The courts' reasons were sufficient for excluding the evidence, and they discussed several important empirical limitations with the evidence. Absent from the discussion at either level, however, were the *conceptual* issues raised by the underlying studies. This is problematic because the conceptual issues also affect inferences that may be drawn from the evidence. And they do so in a way that is even more fundamental than the empirical issues noted by the courts.

The single most important conceptual issue is whether the underlying studies are measuring lies.[47] And, for the reasons discussed earlier, they are not. This can be shown by examining the primary study submitted by Semrau and discussed by the trial and appellate courts. This study, *Functional MRI Detection of Deception After Committing a Mock Sabotage Crime*, of which Laken (Semrau's expert) is a coauthor, reports impressive results in using fMRI to determine which test subjects participated in a mock crime.[48] Although the study describes "lies" and "lying" behavior by participants in each group, the participants were not in fact

[44] Ibid. at 516. [45] Ibid.

[46] Ibid. at note 12 ("Though we are not at that point today, we recognize that as science moves forward the balancing of Rule 403 may well lean toward finding the probative evidence for some advancing technology is sufficient").

[47] The studies also raise the other conceptual issue discussed earlier – namely, the mistaken equation of lies with deception. Laken testified, for example, that "A lie is the intentional act of deceit." Semrau, Hearing Transcript at 159. For the reasons discussed earlier, this conception is underinclusive with regard to lying. See note 23.

[48] Kozel, Johnson et al., 2009. As discussed earlier, see note 16 and accompanying text; the study reports a 100 percent success rate in identifying the participants in a mock crime (but a 33 percent rate in identifying the "no crime" participants).

lying, and the studies were not measuring brain activity during lying behavior. We need to look at the details of experiment to see why.

Participants were randomly divided into two groups: "mock crime" and "no crime."[49] Those in the mock-crime group were given instructions in how to carry out "a mock crime." These instructions told participants to pick up an envelope, "destroy evidence" by breaking a CD, and "steal" a second CD and bring it to the researcher giving the instructions. For those who successfully completed this portion of the study, "[t]hroughout the remainder of the study, the participants were to report that they went to pick up the envelope but did nothing else (i.e., they were to lie about the mock-crime task)."[50] As the authors explain: "any deviation from that would result in the subject being excluded from the study."[51] Those in the "no-crime" group were also instructed to "lie." They were told that they would need an alibi for the time a CD was being destroyed and that they should answer that they were retrieving an envelope in another building at the time. For those in the no-crime group, "they were to answer as if they picked up the envelope but did not damage the CD or visit the crime room."[52] As the authors explain: "[t]he purpose of having the participants lie about picking up the envelope was to have questions during the fMRI evaluation in which participants in this group were lying."[53]

What, exactly, is the problem with this experimental setup? The problem is that lying requires more than simply saying something believed to be false. It also has a normative component. Lying takes place when the speaker communicates something believed to be false in a context where there is a norm against doing so. Violating this norm is what makes prototypical cases of lying constitute "lies" and not other performative acts (such as jokes, game playing, acting, and so on). Importantly, there is no such norm in play during the experiments. Indeed, the participants were instructed to say false things. If anything, the operative norm during the experiment was that they should say false things. In doing so, they were following the rules of the experiment and would be excluded from the study if they failed to do so.[54] The behavior being measured in the experiments more closely resembled game playing, role playing, or acting rather than genuine cases of lying.

49 Ibid. at 221–22. 50 Ibid. at 222. 51 Ibid. 52 Ibid. 53 Ibid.

54 Ibid. Indeed, under this setup, actually telling the truth when they were instructed to say falsehoods would have been closer to real lying.

This mismatch between the studies and actual lying limits the ability to draw inferences about whether behavior in real-world cases constitutes lying. Because the studies were not measuring lying, the relationship between that data and actual lies is essentially unknown. The different context between the studies and the statements by Semrau – his knowingly false statements about his billing practices would be lies – provides another limitation on using the former to drawn inferences about Semrau. In addition to the several empirical limitations discussed by the courts, the fact that the studies[55] were not even measuring lies further limits their utility for real-world settings. Moreover, this conceptual limitation cuts deep. It suggests that the empirical results of such studies do not provide support for additional research into lying; rather, they are built on a conceptual mistake that runs to their foundation. Studies that purport to illuminate lying need to either measure actual lying or provide some reason to think that what is being measured does correlate with lying.

Recent Studies

The conceptual problems that we identified in *Minds, Brains, and Law* occur throughout the fMRI studies on lie detection that had been published at that time. Since we surveyed the literature for our book, the conceptual problems have persisted. This part discusses five examples, focusing on the most important conceptual issue: the criteria for what constitutes a lie and whether the studies are in fact measuring lies.[56]

1. Ofen et al. (2016)[57] – This study purports to measure brain activity correlated with different types of lies.[58] The experiment distinguishes between two different types of false statements: those about past experiences (episodic memories) and those about personal beliefs.

[55] The other studies relied on by Laken and his colleagues involved the same experimental conditions. See Kozel, 2005; F. Andrew Kozel et al., "Replication of Functional MRI Detection of Deception" (2009) 2 *Open Forensic Science Journal* 6.

[56] As with prior studies, the newer studies also continue to conflate lying and deception.

[57] Noa Ofen et al., "Neural Correlates of Deception: Lying about Past Events and Personal Beliefs" (2016) *Social Cognitive and Affective Neuroscience* 1, available at: doi: 10.1093/scan/nsw151.

[58] The study defines a lie as a false statement with an intent to deceive. Ibid. at 4.

The study reports that some brain regions exhibited increased activation for both types of false statements (in the frontal and parietal cortex), but some regions showed differences between the two types of statements (right temporal pole, precuneus, and right amygdala).[59] Based on their experiments, the authors conclude: "[t]he findings here suggest that the brain signature of a lie is influenced by the type of knowledge one is lying about."[60] But were the participants lying? And therefore were the researchers measuring the "brain signature of a lie"? It does not appear so. The participants were instructed to either answer truthfully or falsely to questions about experiences (e.g., "Have you ever ridden a horse?") or opinions (e.g., "Do you believe the death penalty is justifiable?").[61] In this context, they were not violating a norm of communication in speaking falsely. They were instructed to do so; it was a rule of the game that they were playing in the scanner.

2. Jiang et al. (2015)[62] – This study created "whole-brain functional connectivity networks" for each participant and then used a "multivariate pattern analysis approach . . . to distinguish lie-telling from truth-telling."[63] The study reports that "lie-telling" could be differentiated from "truth-telling" with an accuracy rate of 82.81 percent (85.94 percent for lies, and 79.69 percent for truthful statements).[64] The experiment, however, was not measuring lying. Participants in the study were instructed to select three pictures from a group of ten and were then asked during scanning whether a presented picture was one they had selected. During the "truth condition," participants were "required to give accurate, honest responses."[65] During the "lie condition, the participants were required to devise a strategy to deceive others, with the goal of lying skillfully and avoiding detection."[66] Once again, the norm against communicating something false does not apply in this setting, and participants likely did not believe such a norm applied (they were explicitly told otherwise). Thus, the participants were not lying.

[59] Ibid. at 1. [60] Ibid. at 28.

[61] Ibid. at 6. If a participant had intentionally answered contrary to the instruction, then this would better match the normative component of real lying, even if the participant answered truthfully (when instructed to answer falsely).

[62] Weixiong Jiang et al., "Decoding the Processing of Lying Using Functional Connectivity MRI" (2015) 11 *Behavioral and Brain Functions* 1.

[63] The study also equates lying with deception. Ibid. at 1, 6. [64] Ibid. at 4.

[65] Ibid. at 2. [66] Ibid.

3. Sip (2013)[67] – This study explored whether participants' own beliefs about whether their lies could be detected would affect the underlying brain regions involved in producing such lies. Participants were instructed to commit a mock crime (taking earphones or a memory stick from one of two rooms). During the scanning phase, the participants were introduced to an "interrogator" and also hooked up to a fake "lie-detector" machine.[68] They were told the "lie detector" would allow the interrogator "to discriminate between honest and deceptive responses," but that it would be turned on for only half of the session (and they would be told when).[69] This would allow the researchers to examine differences in responses when the participants believed their lies could be detected and when they were not being analyzed. The authors report "a significant interaction between the effects of deception and belief in the left temporal pole and right hippocampus/parahippocampal gyrus ... when participants believed their false claims could be detected, but not when they believed the lie detector was switched-off."[70] Unlike the preceding studies, the participants were not "explicitly instructed to produce false statements"; however, they were instructed that if, after the interrogation, "the interrogator could not tell whether they had taken an object, then they would get to keep the object."[71] Despite this experimental wrinkle (i.e., not telling subjects to say false things), the participants nevertheless do not appear to be lying. Telling participants that they their goal is to trick the interrogator and the "lie detector" plainly conveys to the participants that they are expected to say false things. Or at least that there is nothing normatively problematic about them doing so, unlike in genuine cases of lying. The findings in this study about the participants' *beliefs* about lie detection are certainly relevant and interesting, but the experiment is not measuring lying.
4. Jiang et al. (2013)[72] – This study explored deception among participants with sufficiently high scores on a diagnostic questionnaire for antisocial personality disorder. The study reported increased activation

[67] Kamila E. Sip et al., "When Pinocchio's Nose Does Not Grow: Belief Regarding Lie-Detectability Modulates Production of Deception" (2013) 7 *Frontiers in Human Neuroscience* 1.

[68] Ibid. at 3. This study also equates lying and deception. [69] Ibid. [70] Ibid. at 1.

[71] Ibid. at 3–4.

[72] W. Jiang et al., "A Functional MRI Study of Deception among Offenders with Antisocial Personality Disorder" (2013) 244 *Neuroscience* 90. The study also equates lying with detection.

in several brain regions during "lie-telling."[73] This experiment, however, employed the same select-three-pictures paradigm discussed earlier (with instructions to answer truthfully when instructed to do so and to "lie" when instructed to do so).[74] Thus, the experiment was not measuring lying, even in the "lie-telling" condition.

5. Cui et al. (2013)[75] – This study instructed subjects to participate in a "mock murder." The participation consisted of creating a story with several details about a fictitious murder (for example, identity of victim, weapon used). Other subjects (the "innocents") did not participate in constructing a story about a murder. Both groups were scanned and asked details about the "murder," with innocents instructed to honestly deny knowledge and the "mock murders" instructed to lie.[76] The authors report different activation in brain areas in the two groups. Specifically, "the right VLPF [ventral lateral prefrontal] contributed the most to diagnosing the participants' real identities" – "Based on the activity of this area, 81.25 percent of the 'murderers' and 81.25 percent of the 'innocents' were correctly diagnosed."[77] Subjects in this study also appear not to be lying. As with the other studies, participants do not appear to be violating a norm against communicating something believed to be false – they were instructed to do precisely that – and thus their false statements are not lies.

The purpose in critiquing these studies – and with the previous ones discussed in *Minds, Brains, and Law* – is to illustrate the importance of conceptual issues for law and neuroscience. The empirical limitations on using neuroscientific evidence in court are real and have received important treatment from scholars in a variety of fields. The success of empirical investigations, however, also depends on clear conceptual presuppositions underlying the investigations. Philosophical attention to these

[73] Ibid. at 96. ("The pattern of brain activity in the frontal-anterior cingulated-parietal regions associated with lying that was observed in our study is largely consistent with the findings of previous fMRI deception studies.")

[74] Ibid. at 93. See notes 62–66 and accompanying text.

[75] Qian Cui et al., "Detection of Deception Based on fMRI Activation Patterns Underlying the Production of a Deceptive Response and Receiving Feedback about the Success of the Deception after a Mock Murder Crime" (2013) *Social Cognitive and Affective Neuroscience* 1.

[76] Ibid. at 2–3.

[77] Ibid. at 7. The study also reports increased accuracy rates when subjects were given feedback about whether their prior answer were judged to be deceptive or not. Ibid. (93.75 percent of the "murderers" and 87.5 percent of the "innocents").

presuppositions can help to clarify the issues and correct mistaken assumptions. This suggests an important role for philosophy to play in neurolaw, not only for issues involving high-level theoretical issues (such as free will and criminal responsibility) but also at the level of evidence and proof on the ground. The conceptual issues involved with fMRI lie detection are one such example. In critiquing the conceptual underpinnings of some of the scientific literature, the aim has not been to challenge the *possibility* of brain-based lie detection. Rather, the aim has been to ensure that any evidence that is produced for legal purposes is probative regarding something the law cares about – in this instance, lying rather than other types of behavior.

7

Brain-Based Lie Detection and the Mereological Fallacy: Reasons for Optimism

JOHN DANAHER

There has been much hype about the implications of contemporary developments in neuroscience for the law. Pardo and Patterson are skeptical of this hype.[1] They argue that a good deal of the hype stems from simple philosophical errors and conceptual confusions. In the course of this critique, they offer particular objections to the forensic use of brain-based lie-detection methods. Although agreeing with the authors about the need for skepticism and conceptual clarity, this chapter argues that they get things wrong when it comes to their skepticism of brain-based lie detection. This is for three reasons. First, in their critique they focus too heavily on the problems associated with the more speculative and less empirically grounded fMRI-based methods and not enough on the more robustly grounded EEG-based methods. Second, when focus is switched to these methods, their main philosophical critique of the use of neuroscience in law – the *neurolaw mereological fallacy* – has much less bite. And third, they neglect to address the merits of brain-based lie-detection methods *relative to* existing methods for inferring what a witness does or does not believe. When these three critiques are factored in, the future looks brighter for this particular use of neuroscience in law.

Introduction

There is no doubt that the legal implications of neuroscience have been overhyped and oversold. Michael Pardo and Dennis Patterson's *Minds, Brains, and Law* is a useful corrective to this hype. In the book, Pardo and Patterson make a passionate plea for conceptual clarity when it comes to the interpretation and application of neuroscientific evidence in the law.

[1] Pardo, Michael, and Patterson, Dennis, *Minds, Brains and Law* (Oxford: Oxford University Press, 2013).

I am sympathetic to their aims. I too am keen to downplay the neurolaw hype.[2] I too think that there is much conceptual and philosophical confusion in the literature. That said, one area of neurolaw about which I am slightly more bullish is the potential use of brain-based lie-detection methods in forensic settings.[3] But let me clarify. I am not bullish about 'lie detection' per se, but rather about EEG-based *recognition detection tests* or *concealed information tests.* In particular, I am bullish about tests based on the P300 brainwave and its ability to link suspects to incriminating evidence.

Pardo and Patterson do not seem to share my optimism. Consistent with their overarching thesis, they argue that the forensic usefulness of this technology has been oversold and overhyped, that the boosters of the technology are guilty of serious conceptual confusions, and that this undermines their claims about its practical utility. Although they are not completely dismissive, they are arch-skeptics.

In this chapter I want to push back against this skepticism. I want to argue that Pardo and Patterson's worries about these tests are misplaced or readily addressed. There are three main reasons for this. First, in their discussion of brain-based lie detection, they focus too much on fMRI-based control question tests (CQTs), and not enough on EEG-based concealed information tests (CITs). As a result of this, they fail to properly engage with the most interesting and forensically useful current work being done on brain-reading technology. Second, when the focus switches to this alternative type of brain-reading technology, their central philosophical critique of neurolaw (the 'mereological fallacy' argument) lacks purchase. And third, one of the main reasons why it lacks purchase, and one of the main reasons for optimism about this technology, is that this technology has the potential to improve the forensic position *relative* to the current status quo. When you factor in the epistemological limitations of the current status quo, you see that the technology – limited

[2] See Danaher, John, 'The Future of Brain-Based Lie Detection and the Admissibility of Scientific Evidence' (2011) *Irish Criminal Law Journal* 21(4): 67–76 and Danaher, John, 'Scientific Evidence and the Criminal Law: Lessons from Brain-Based Lie Detection' (2010) *Judicial Studies Institute Journal* 10(1): 94.

[3] Danaher, J. 'The Comparative Advantages of Brain-Based Lie Detection: The P300 Concealed Information Test and Pre-Trial Bargaining' (2015a) *International Journal of Evidence and Proof* 19(1): 52–66 and Danaher J., 'Responsible Innovation in Social Epistemic Systems: The P300 Memory Detection Test and the Legal Trial' in Van den Hoven et al. (eds.), *Responsible Innovation Volume II: Concepts, Approaches, Applications* (Dordrecht: Springer, 2015b).

and flawed though it may be – could resolve or help overcome those epistemological limitations. I develop this argument with a particular example (one that I have used in the past): the ability of the P300 to address the 'innocence problem' in pretrial plea-bargaining.[4]

The chapter proceeds as follows: I will first review the basic techniques of brain-based lie detection, developing a two-by-two matrix for distinguishing the different methods. At this point I will introduce the first of my criticisms, suggesting that Pardo and Patterson unhelpfully prioritize one type of test in their analysis. Moving on from this, I will summarize Pardo and Patterson's main conceptual objection to brain-based lie detection (the 'Mereological Fallacy' argument) and see how it applies to the different testing protocols. I will then develop the two further stages of my critique, suggesting that this objection has less bite when it comes to the P300 concealed information test and that the propensity toward status quo bias must be borne in mind when assessing the merits of that technology.

Brain-Based Lie Detection: A Guide to the Territory

Working out what people know, think and believe is important to the law. Many legal liability tests make appeals to states of mind (intention, knowledge, will and so on), and many forensic inquiries are dedicated to working out what people saw, thought and believed at some historic moment. There is nothing particularly unusual about law in this regard. In our everyday lives we frequently want to know what other people think, believe and know. It is essential to social coordination and discourse.

The most direct and obvious way to find out what people know or think or believe is to simply ask them. The problem in forensic settings is that this direct route is often misleading. Given what is at stake, people have an incentive to deceive or mislead the court or tribunal with respect to what they *really* know, think and believe. Indeed, sometimes they have a legally protected right not to reveal what they really know, think and believe. On top of this, individuals who willingly present testimony to courts and tribunals can misremember or misstate the truth, or could be unable to reveal their state of mind in a legally relevant manner.

The great promise of lie-detection tests is their ability to cut through some of these epistemological problems. If we can use a physiological

[4] Danaher, ibid., 2015a.

measure (e.g., sweaty palms, blood-enriched brain regions) to detect what someone is *really* thinking, we can obviate the need to ask them or, at the very least, we can weigh what the physiological measure is telling us against what they are telling us and perhaps favor the former over the latter. Traditional variants of the lie detector do this by measuring variations in the physiological activity in the autonomic nervous system (e.g., galvanic skin response, systolic blood pressure). The more modern brain-based variants go directly to what we believe is the physiological correlate of the mind: the brain.

When debating the merits of these brain-based variants of lie detection, it is important to distinguish between two things: (1) the scanning technology and (2) the testing protocol. The *scanning technology* is what provides us with data about an individual's brain activity.[5] There are currently two main technologies used in this regard. Functional magnetic resonance imaging (fMRI) is used to track variations in the flow of oxygenated blood across different brain regions.[6] This is typically thought to be a good proxy measure for underlying brain activity. The data from fMRI scans can be processed into images of activation patterns and then overlaid onto a visual representation of the brain. Such images have become nearly ubiquitous in popular representations of neuroscience. Electro-encephalographic imaging (EEG) is slightly different and less 'sexy'. It tracks variations in electrical activity across the scalp producing a readout that represents this activity in a wave-like pattern. This too is thought to be a good measure of underlying brain activity, though the measure is cruder than that provided by fMRI ('cruder' in the sense that it is less capable of being localized to a specific subregion of the brain).

The *testing protocol* is how the data provided by the scanning technology is used to find out something interesting about what the test subject really knows, thinks and believes. In a *control question test* (CQT) the data being read from the scanning technology is used to make

[5] See Jones, O., Buckholtz, J., Schall, J. and Marois, R. 'Brain-Imaging for Legal Thinkers: A Guide for the Perplexed' (2009) *Stanford Technology Law Review* 5 – available at: https://journals.law.stanford.edu/sites/default/files/stanford-technology-law-review/online/jones-brain-imaging.pdf (accessed 19/10/2016).

[6] See Kolb, B. Wishaw, I and Campbell Tesky, G. *An Introduction to Brain and Behavior*, 5th ed. (New York: Worth Publishers, 2016); Logothetis, N. (2008). What We Can and Cannot Do with fMRI. *Nature* 453: 869–878; Jezzard, P., Matthews, P., and Smith, S. *Functional MRI: An Introduction to Methods* (Oxford: Oxford University Press, 2003).

inferences about whether a test subject is lying or being deceitful. The CQT is the testing protocol most commonly associated with the traditional polygraph lie detector and will be familiar to people from pop cultural representations of the lie-detector test. A typical CQT is set up in the following manner. A subject is hooked up to the scanning technology as part of an investigation into their knowledge or understanding of some incident. The subject is then asked a series of questions, some of which relate to that incident. These questions usually fall into three categories: (i) *control questions*, i.e., general questions that most people would be inclined to lie about; (ii) *relevant questions*, i.e., those specifically relevant to the incident under investigation; and (iii) *irrelevant questions*, i.e., those with no bearing on the matter under investigation. The theory underlying the CQT is that deceptive test subjects will exhibit a consistently larger physiological response to relevant questions than to control questions; innocent suspects will do the reverse. The belief then is that you can infer deceptiveness from this difference in response profiles. Despite its popularity, the CQT has been repeatedly criticized for being insufficiently scientific, easily manipulated by the tester and often adopted simply as an interrogation prop to force a confession.[7]

The CQT is to be contrasted with the *concealed information test* (CIT), which does not try to assess whether a test subject is being deceptive.[8] Instead, it tries to assess whether they, or more correctly their brain,

[7] See Ben-Shakhar, G. 'A Critical Review of the Control Questions Test' in M. Kleiner (ed.), *Handbook of Polygraph Testing* (Academic Press: London, 2002); Furedy, J. J. 'The "Control" Question "Test" (CQT) Polygrapher's Dilemma: Logico-Ethical Considerations for Psychophysiological Practitioners and Researchers' (1993) 15 *International Journal of Psychophysiology* 263–7; Furedy, J. and Heselgrave, R. The Validity of the Lie Detector: A Psychophysiological Perspective (1988) 15 *Criminal Justice and Behavior* 219–246. National Academies of the Sciences Report, *The Polygraph and Lie Detection* (Washington, DC: National Academies Press, 2003).

[8] CITs can sometimes be used to infer deceptiveness because the test subject will be asked to give a response to the stimuli presented. The response may be 'I recognize this' or 'I don't recognize this'. If there is a discrepancy between the subject's response and the signal from their brain, you can infer deceptiveness. Whether forcing such a response helps improve the accuracy of the test is a matter of debate among researchers see Meegan, D. Neuroimaging Techniques for Memory Detection' (2008) 8 *American Journal of Bioethics* 9–20; and Rosenfeld, J. P., Hu, X., Labkovsky, E., Meixner, J. and Winograd, M., 'Review of Recent Studies and Issues Regarding the P300-Based Complex Trial Protocol for Detection of Concealed Information' (2013) 90 *International Journal of Psychophysiology* 118–34.

recognizes certain information.[9] Although protocols for the CIT can vary, in broad outline they work like this: the test subject is hooked up to some scanning technology. They are then presented with different kinds of stimuli (e.g., images or words), some of which are relevant to an incident under investigation (usually called the 'probe' stimuli), some of which are not (but which can belong to a similar category of stimuli). A heightened physiological response to the probe stimuli suggests that the subject recognizes the information. It is easy to see how this could form the basis of a forensically useful test. For example, in the case of a criminal investigation a suspect could be tested for their recognition of crime-relevant details, e.g., murder weapon, venue, victim's dress and so on. Because the information in question will usually be tied to a crime scene, the test is sometimes referred to as the *guilty knowledge test.* But this name is unfortunate and should be avoided: the test does not prove guilt or even knowledge – two critical points to which I shall return later on. At best, it proves recognition of information. Further inferences must be made to prove that a suspect has guilty knowledge. This is why the term 'concealed information test' has caught on. But calling it the 'concealed' information test is not great either because the suspect may or may not be 'concealing' the information in question. For these reasons, I tend to prefer calling it something like a memory detection test or, better, a *recognition detection test*, but concealed information test is the norm within the literature, so I will stick with that[10]

With two main scanning technologies[11] and two main test protocols, we can think about the space of possible brain-based lie-detection methods in terms of the following two-by-two matrix.

The matrix in Figure 7.1 represents the logical space of possible tests. Whether or not tests exist in all four quadrants depends on whether there are (a) physiological signals that can be detected by the relevant scanning technologies that (b) can, in turn, form the basis for a CQT and CIT. Although EEG-based CQTs and fMRI-based CITs are physiologically possible (indeed, some research has been done into the viability of an

[9] For an introduction see: Verschuere, B., Ben-Shakhar, G., and Meijer, E. *Memory Detection: Theory and Application of the Concealed Information Test* (Cambridge: Cambridge University Press, 2011).

[10] cf. Meegan, and Verschuere et al., who go with 'memory detection'. See Meegan (2008) and Verschuere et al., ibid.

[11] There are, of course, other kinds of brain scanning technologies, but they have not yet taken off as possible bases for lie detection tests.

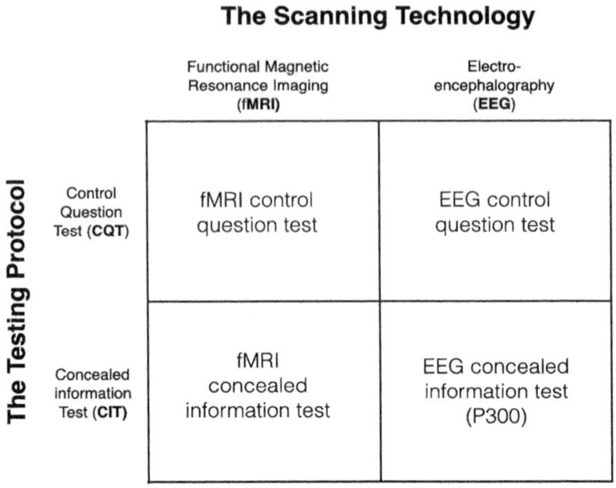

Figure 7.1 Possible forms of brain-based lie detection

fMRI CIT[12]), it is in the other two quadrants that most of the research activity has been focused. Thus, when we talk about brain-based lie detection, we are really only talking about two types of test: fMRI CQTs, and EEG CITs. Both have been the subject of media attention and forensic consideration.

A number of fMRI studies in the early 2000s located potential brain markers of deception.[13] These studies suggested that an fMRI CQT was viable: if a certain pattern of activity was detected by the fMRI when someone answered a question it could indicate that they were being deceptive. In the wake of these studies, two companies were set up in the United States that offered to administer fMRI CQTs for forensic purposes: No Lie MRI[14] and Cephos Corp[15] (the former still seems to be in existence, but the latter has shifted its attention to DNA testing).

12 See Gamer, M. 'Detecting of Deception and Concealed Information Using Neuroimaging Techniques' in Verschuere, Ben-Shakar and Meijir (eds.) (2011).

13 The scientific evidence in favor of an fMRI CQT is ably reviewed in Pardo and Patterson (2013), 82–86.

14 For information see www.noliemri.com (visited 19/10/2016).

15 For information see http://cephosdna.com (visited 19/10/2016) – there is little information about the fMRI CQT on that page now, but the company founder, Dr Steven Laken, has presented fMRI evidence in court on previous occasions (see the *Semrau* decision, next footnote).

The results of fMRI CQTs were even considered for admission into court in a number of cases but were refused in all instances.[16] Despite these setbacks, it should be noted that experimental tests of the fMRI CQT suggest that it has an accuracy rate somewhere in the 80 per cent range, which seems like it could be forensically valuable (probative) in certain contexts (see Schauer, 2010 and 2012 on this point).

The EEG CIT has a much longer history. It is premised on the detectability of a particular kind of brainwave known as the P300 (because it appears roughly 300ms after the presentation of a stimulus to the test subject). The existence of the P300 has been known for over a half-century.[17] There is a rich research literature that tries to understand exactly what it signifies. The leading theory is that the P300 is present whenever a test subject views a rare and meaningful (i.e., recognizable to them) stimulus. Because of this it is thought to provide a promising basis for a CIT – the idea being that if you present a test subject with information linked to a crime (or other incident), the P300 will be detected if they (or, rather, *their brain*) recognizes the information.

Early mock-crime tests of the P300-based CIT showed considerable promise. These tests were conducted by Lawrence Farwell and Emmanuel Donchin[18] and suggested that very high accuracy rates ($>$ 90 per cent) could be achieved with the P300 measure, with errors tending in favor of false negatives rather than false positives. In other words, the test was more likely to falsely label someone as not recognizing crime-relevant information than to falsely label them as recognizing the information – advantageous given the oft-cited normative preferences of the criminal law for exculpation over inculpation. Following these initial experiments, Farwell developed his own, patented version of the test (sometimes referred to as 'brain fingerprinting') and received some reasonably high-profile media exposure for his efforts to get it accepted by the courts.

[16] *United States v Semrau*, No. 07–10074 M1/P (WD Tenn. May 31, 2010); and *Wilson v Corestaff Services LP* (2010) 900 NYS 2d 639. The evidence was omitted in the former case because it failed to pass the *Daubert* test for admissibility. The judge expressed particular concern about the error rates for the test. It was rejected in the latter because it failed to pass the *Frye* test for admissibility and was proffered in relation to a matter well within the competence of the jury.

[17] For the history see, Rosenfeld, J. P. 'P300 in Detecting Concealed Information' in Verschuere, Ben-Shakar and Meijir (eds.) (2011); and Rosenfeld et al. (2013).

[18] Farwell, L. and Donchin, E. 'The Truth Will Out: Interrogative Polygraphy ('Lie Detection') with Event Related Potentials' (1991) 28 *Psychophysiology* 531–47.

These efforts were largely unsuccessful.[19] In subsequent years, Farwell involved himself in classified military and security service tests of his technique, before re-emerging into the academic world to argue for the strengths of his version of the P300 CIT.[20] But Farwell is far from the only person suggesting that the P300 CIT has forensic uses. Other researchers have conducted scores of empirical studies of the test, highlighting its empirical strengths and weaknesses.[21]

The First Critique: Overemphasizing the fMRI CQT?

Pardo and Patterson's critique of brain-based lie detection has conceptual, empirical and practical dimensions. Their main goal is to emphasize the conceptual confusions underlying these tests and to suggest that these conceptual confusions have serious repercussions for the practical utility of the tests. Thus, their conceptual case against brain-based lie detection is the most important aspect of their critique. I will discuss this in the next section. In the meantime, however, I want to dwell on the empirical aspect of their critique. This seems appropriate because they too focus on the empirical before they turn to the conceptual.

Pardo and Patterson are well aware that there are two major types of brain-based lie-detection tests. They discuss both at some length. Furthermore, their discussion shows all the hallmarks of familiarity with the empirical literature. Nevertheless, it is here that the first problem arises. Although Pardo and Patterson are well informed about both tests, their discussion places far more emphasis on the merits and demerits of the fMRI CQTs and, as a result, glosses over or ignores some of the reasons for feeling bullish about the forensic uses of the P300 CIT. This is problematic insofar as it is the latter that I would argue has the most forensic potential, not the former. This is due to both inherent problems

19 See Farwell's company web page, located at www.brainwavescience.com (visited 19/10/16). The evidence was presented to court to assist with exoneration in two death penalty review cases: *Harrington v State of Iowa* 659 NW.2d 509 (Iowa 2003) and *Slaughter v State of Oklahoma* 105 P.3d 832 (Oklahoma 2005).

20 See Farwell, L., Richardson, D. and Richardson, G. 'Brain Fingerprinting: Field Studies Comparing P300-MERMER and P300 Brainwave Responses in the Detection of Concealed Information' (2013) 7 *Cognitive Neurodynamics* 263–99; Farwell, L. Brain Fingerprinting: A Comprehensive Tutorial Review of Detection of Concealed Information with Event-Related Brain Potentials (2012) *Cognitive Neurodynamics* 6: 115–54.

21 For overviews, see Verschuere et al. (2011); Meixner, J., 'Applications of Neuroscience in Criminal Law: Legal and Methodological Issues' (2015) 15 *Current Neurology and Neuroscience Reports* and Rosenfeld et al. (2013).

in the CQT testing protocol and the weight and complexity of the scientific evidence in favor of the P300 CIT. Allow me to explain.

Pardo and Patterson highlight several limitations in the current empirical support for both tests[22]). The essence of their complaints will be familiar to anyone who has researched this topic. Although they acknowledge that the empirical studies of the accuracy of fMRI CQTs and P300 CITs are 'mixed but overall impressive', they nevertheless worry that '[the] results [of those studies] are based on highly artificial and tightly controlled testing conditions; therefore the validity and reliability of the technique when applied to more complex, "real world" situations is unknown'.[23] They then go on to list a number of more-specific empirical shortcomings that will need to be addressed if the tests are to have real forensic utility. Eight problems in total are listed,[24] including problems with how raw data gets converted into images, the small sample size in most tests, the lack of diversity in test subjects, the fact that the time lag between when experimental subjects acquire information and when they are tested is much shorter in empirical studies than it would be in real-world investigations, the failure to properly distinguish between different kinds of lie and deception, the low stakes of the test environment, the correlation between the brain activity and other behavioral and mental activity (i.e., confounds), and finally, the susceptibility of these tests to countermeasures (i.e., efforts to fool or thwart the test).

Although each of these empirical limitations is worthy of our attention, several of them apply primarily to fMRI CQTs, not to the P300 CIT. Furthermore, in their presentation of these limitations, Pardo and Patterson fail to do justice to the ongoing research into the P300 CIT that tries to address them. This a point worth belaboring. There is undoubtedly a danger that brain-based lie detection will be foisted upon an unwitting and unprepared legal community before it is truly ready for forensic primetime. This danger has already materialized. As noted earlier, the fMRI CQT and the P300 CIT already have (or have had) their forensic boosters: No Lie MRI and Cephos in the case of fMRI CQTs and Lawrence Farwell in the case of the P300 CIT. These boosters have realized the commercial potential for the technology and are eager to capitalize on it. But they have done so prematurely, before all the important empirical kinks have been ironed out. Given this real and present danger, it is absolutely vital for the scientific and

[22] Pardo and Patterson (2013), 93–94 and 97–99. [23] Ibid. at 93. [24] Ibid. at 98–99.

legal communities to be alerted to those empirical kinks and subject any forensic uses of these tests to appropriate critical scrutiny. But sensitivity to this danger should not prevent us from appreciating that there are more cautious scientific proponents of these technologies that are sensitive to their empirical shortcomings and are doing the painstaking empirical work that is needed to address these shortcomings. To simply repeat the standard empirical concerns in ignorance of this work is to do a disservice to the debate about the forensic use of the technology.

This seems to be what is happening in Pardo and Patterson's critique of the P300 CIT. They highlight the case of Lawrence Farwell, and it is fair to say that he pushed the forensic use of this test far too soon. But Farwell's efforts in this regard have been roundly condemned by his peers in the P300 research community, and the theoretical/conceptual confusions underlying his public presentation of the test have been highlighted by those researchers.[25] Some of this condemnation has been strongly worded. For example, one review of work done by Farwell concluded

> By selectively dismissing relevant data, presenting conference abstracts as published data, and most worrisome, deliberately duplicating participants and studies, he misrepresents the scientific status of brain fingerprinting. Thus, [Farwell] violates some of the cherished canons of science and if [he] is, as he claims to be, a "brain fingerprinting scientist" he should feel obligated to retract the article.[26]

These other researchers are keen to correct for Farwell's mistakes and try to address the shortcomings highlighted by Pardo and Patterson. The work being done by the Rosenfeld Lab at Northwestern University is a clear example of this.[27] Researchers at that lab have published dozens of studies in the past 15 years that try to improve the ecological validity of the tests. They have done so in several ways, for instance, by examining how incidentally acquired information can be detected by the test, exploring how long a time lag between an incident and a test there can be before there is some falloff in accuracy, and examining the susceptibility of the

[25] See Meijer, E. Ben-Shakhar, G., Verschuere, B. and Donchin, E. 'A Comment on Farwell (2012): Brain Fingerprinting: A Comprehensive Tutorial Review of Detection of Concealed Information with Event-Related Brain Potentials (2013) 7 *Cognitive Neurodynamics* 155–8; and Rosenfeld, J. P. 'Brain Fingerprinting: A Critical Analysis' (2005) 4 *Scientific Review of Mental Health Practice* 20–37.

[26] Ibid. at 158.

[27] See http://groups.psych.northwestern.edu/rosenfeld/home.html (accessed 22/3/2017).

P300 CIT to countermeasures. In the process, they have discovered that incidentally acquired information is still well detected by the test,[28] that the test is reliable up to a month after an incident has occurred,[29] that the test is vulnerable to countermeasures,[30] but that there are techniques that can be adopted to make it more resistant and more accurate.[31] They have also constructed more realistic mock-crime and real-world tests, but still insist that extensive field-testing of the P300 CIT is needed before it can be used more widely.[32] Although Pardo and Patterson seem to be aware of some of this work (and although some of it was published after their book) they don't explore the depths and nuances of it. If they did, I think they might be more inclined to share my optimism.

In addition to this, there are two minor points worth noting about the P300 CIT that gives it advantages over the fMRI CQT. Again, these are not explored by Pardo and Patterson. The first is that by relying on the CIT test format, the P300 CIT addresses some of the criticisms that have long been leveled at the CQT, namely: that it is little more than an interrogation prop designed to force a confession and that it is easily manipulated by the tester. The second is that although there are studies suggesting that the P300 has a low overall accuracy rate in some more

[28] See Meixner, J. and Rosenfeld, J. P. 'Detecting Knowledge of Incidentally Acquired, Real-World Memories Using a P300-Based Concealed-Information Test' (2015) 25 *Psychological Science* 1994–2005; and Winograd, M. and Rosenfeld, J. P. 'Mock Crime Application of the Complex Trial Protocol P300-Based Concealed Information Test' (2011) 48 *Psychophysiology* 155–61.

[29] Hu, X. and Rosenfeld, J. P. (2012). 'Combining the P300-Complex Trial-Based Concealed Information Test and the Reaction Time-Based Autobiographical Implicit Association Test in Concealed Memory Detection' (2012) 49 *Psychophysiology* 1090–100.

[30] Rosenfeld, J. P., Soskins, M., Bosh, G. and Ryan, A. 'Simple, Effective Countermeasures to P300-Based Tests of Detection of Concealed Information' (2004) 41 *Psychophysiology* 205–9.

[31] See Hu, X., Hegeman, D., Landry, E., and Rosenfeld, J. P. 'Increasing the Number of Irrelevant Stimuli Increases Ability to Detect Countermeasures to the P300-Based Complex Trial Protocol for Concealed Information Detection' (2011) 49 *Psychophysiology* 85–95; Hu, X., Pornpattananangkul, N., and Rosenfeld, J. P. 'N200 and P300 as Orthogonal and Integrable Indicators of Distinct Awareness and Recognition Processes in Memory Detection' (2013) 50 *Psychophysiology* 454–64.; Meixner, J. and Rosenfeld, J. P. 'Countermeasure Mechanisms in a P300-Based Concealed Information Test' (2010) 47 *Psychophysiology* 57–65; Rosenfeld, J. P., Labkovsky, E., Winograd, M., Ming, A. L., Vandenboom, C., and Chedid, E. 'The Complex Trial Protocol (CTP): A New, Countermeasure-Resistant, Accurate, P300-Based Method for the Detection of Concealed Information' (2008) 45 *Psychophysiology* 906–19.

[32] See Meixner, J. 'Liar, Liar, Jury's the Trier' (2012) 106 *Northwestern Law Review* 1451–88; Rosenfeld et al. (2013) and Meixner, J. (2015).

realistic experimental tests,[33] it is important to note that in these tests the results are usually biased in favor of false negative rather than false positive. In other words, the test is more likely to falsely exonerate a test subject than falsely incriminate them. This becomes particularly significant when we consider the comparative advantages argument that I discuss later.

The Mereological Fallacy Argument

Enough with the empirical concerns. More important to Pardo and Patterson's project is the conceptual confusion that they think underlies much of the ongoing conversation about brain-based lie detection and its forensic uses. This is the heart of their main argument: the 'mereological fallacy' argument. I will need to explain what this is, and how it applies to brain-based lie detection tests, before I can proceed to my final two criticisms.

Mereology is the study of part–whole relations, so as you might gather, the mereological fallacy arises from a failure to appreciate the difference between a whole and one of its parts. For the sake of clarity, let us distinguish between two versions of the mereological fallacy. The first, more general one, can be defined like this:

> *The General Mereological Fallacy*: Arises whenever you ascribe properties that are rightly ascribed to a whole to some part or subpart of that whole. For example, applying the predicate 'fast' to a runner's leg, rather than to the runner.

The second, more specific one, can be defined like this:

The Neurolaw Mereological Fallacy: Arises whenever a neurolaw proponent ascribes behavioral or person-level properties to a state of the brain. (In other words, whenever they assume that a brain state is constitutive of or equivalent to a behavioral or person-level state). For example, applying the predicate *wise* to a state of the brain, as opposed to the person whose brain state it is. Similarly applying the predicate *lying* or *knowledge* to a part of the brain as opposed to the person.

This more specific version of the fallacy is the centerpiece of Pardo and Patterson's book. Indeed, their book is effectively one long elaboration of how the neurolaw mereological fallacy arises in various aspects of the

[33] E.g., Mertens and Allens report a less than 50 per cent accuracy rate: Mertens, R. and Allen, J. J. B. 'The Role of Psychophysiology in Forensic Assessments: Deception Detection, ERPs and Virtual Reality Mock Crime Scenarios' (2008) 45 *Psychophysiology* 286–98.

neurolaw literature. In basing their criticism on this fallacy, they are following the work of others.[34]

The neurolaw mereological fallacy provides the basis for Pardo and Patterson's main critique of brain-based lie detection. Though they do not set this critique out with any formality, I think it can be plausibly reconstructed as taking the following form:[35]

1. If it is likely that the use of brain-based lie detection evidence would lead legal actors (lawyers, judges, juries, etc.) to commit the neurolaw mereological fallacy, then we should be [very] cautious about its forensic uses.
2. The use of brain-based lie detection evidence is likely to lead legal actors to commit the neurolaw mereological fallacy.
3. Therefore, we should be (very) cautious about the forensic uses of brain-based lie-detection evidence.

The first premise is the guiding normative assumption. I am not going to challenge it here. I will simply accept it *arguendo*. Nevertheless, one might wonder why we should endorse it. Why is the mereological fallacy so normatively problematic? There are some reasons. The main one is that the law cares about certain concepts. These include the *intentions* of a murder suspect, the *knowledge* of the thief, and the *credibility* or potential *deceptiveness* of the witness. The application of these concepts to real people is what carries all the normative weight in a legal trial. The content of one's intentions and the state of one's knowledge is what separates the criminal from the innocent. The deceptiveness of one's testimony is what renders it probative (or not) in legal decision making. Pardo and Patterson maintain that all these concepts, properly understood, apply to the behavioral or personal level of analysis. For example, they argue that deceptiveness depends on a complex relationship between a person's behavior and the context in which that behavior is performed. To be precise, they argue being deceptive means saying or implying something that one believes to be false, in a social context in which truth-telling is expected or demanded.[36] These behavioral-contextual criteria are what ultimately determine the correct application of the predicate 'deceptive' to an individual.

[34] Bennett, M. R. and Hacker, P. M. S. *The Philosophical Foundations of Neuroscience* (Oxford: Blackwell, 2003).

[35] Patterson and Pardo (2013) at 99–105.

[36] Ibid., 109–10.

If we make a mistake in the application of those predicates, it has significant normative implications. If we deem someone deceptive when, by rights, they are not, then we risk a miscarriage of justice (or something less severe but still malign). The concern that Pardo and Patterson have is that neurolaw will encourage people to make such mistakes. If they start using neurological criteria as the gold standard in the application of normative, behavioral-level predicates like *intention*, *deception* and *knowledge*, then they risk making normative errors. This is why we should be cautious about the use of neuroscientific evidence in the law.

How cautious should we be? That's something I'm not entirely clear about from my reading of Pardo and Patterson's book. They are not completely opposed to the use of brain-based lie detection in the law. Far from it. They think it could, one day, be used to assist legal decision making. But they do urge some level of caution. My sense from their discussion, and from their book as a whole, is that they favor a lot of caution. This is why I have put *very* in brackets in my statement of premise (1).

Moving on then to premise (2), this is the key factual claim about the use of brain-based lie-detection evidence. In its current form it does not discriminate between the P300 CIT and the fMRI CQT. Pardo and Patterson's concern is that evidence drawn from these tests will lead legal actors to confuse the presence of brain signal *X* with the subject's meeting the criteria for the application of a behavioral predicate like 'knowing' or 'intending' or 'deceiving'. In the case of the P300 CIT, the fallacy arises if the detection of the P300 is taken to be equivalent to the detection of a 'knowledge'-state within the subject's brain, instead of merely evidence that can be used to infer that the subject is in the appropriate *behavioral* knowledge state. Similarly, in the case of the fMRI CQT, the fallacy arises if people assume that the detection of a particular pattern of activation indicates that the person is lying or dissembling, instead of merely providing evidence that can be weighed (along with other behavioral evidence) in making such an inference. They are adamant that this could be a serious problem for the law:

> People, not their brains, lie, deceive, know, and remember. It is an instance of the mereological fallacy to suppose otherwise, and the evidential value of neuroscientific evidence is affected by how it is conceived. If lies, knowledge, and intent are mistakenly thought to be identical with particular brain states, then proof of the brain states may (again, mistakenly) appear to provide conclusive or strong proof of the corresponding mental states.

> Indeed those in the grip of this fallacy might even suggest that we attach legal consequences directly to people who have brains in particular neurological states. Once it is recognized, however, that such evidence may have some inductive correlation – but does not provide a sufficient condition to establish the categories at issue – then mistaken inferences to the contrary will be avoided.[37]

Is this problem likely to occur?

The Second Critique: The Mereological Fallacy Is Unlikely or Easily Addressed in the Case of the P300 CIT

Pardo and Patterson's conceptual point is well taken. I quite agree that behavioral and personal-level properties like knowledge, deception, intention and so forth cannot be simply reduced to (or eliminated by) brain states. I also agree that evidence from brain-based lie-detection tests must be understood to provide *inductive* but not *criterial* evidence for those behavioral properties. The critical question then becomes: how likely is it that proponents and users of this technology will make the fallacy? That is a tough question to answer. It will depend on how the evidence is interpreted in the relevant scientific community and how it is presented to and understood by the legal community. I will not address both sides of this equation in this chapter, but I will say something about the interpretation and understanding in the relevant scientific community.

I think Pardo and Patterson present a strong case for thinking that current scientific research into the fMRI CQT is marred by something pretty close to the mereological fallacy. As they point out, lying is a behavioral and contextual phenomenon: it requires saying something you know to be false in a context in which truth-telling is expected. It is almost impossible to reconstruct these behavioral and contextual conditions in an experimental test for an fMRI CQT. Instead, you have to instruct or force some of the subjects to lie. This suggests to Pardo and Patterson[38] that the participants in these tests are not really lying, which severely undermines the inductive inferences one can make from these experiments to real-world cases.

But this problem does not arise in the case of the P300 CIT. This is for two reasons. First, *knowing* or, more correctly, *recognizing* information

[37] Ibid. at 105. [38] Ibid. at 109–10.

does not require similar behavioral or contextual conditions to be met in an experimental test. It is much easier to set up an experimental test that can provide the basis for plausible inductive inferences. Second, it is not clear that the inferences that need to be (and are) made from the results of a P300 CIT come anywhere close to the mereological fallacy. The second reason is the more important one and requires some unpacking.

Pardo and Patterson are clearly worried about the mereological fallacy arising in the case of the P300 CIT. They offer support for their worries by quoting from Lawrence Farwell and Sharon Smith. When describing how the P300 CIT worked, Farwell and Smith once said that the 'brain of the criminal is always there, recording events, in some ways like a video camera'.[39] Hence, they suggested that the P300 CIT directly revealed whether or not the suspect knew the incriminating information. Pardo and Patterson insist that Farwell is committing the fallacy here because he thinks that the state of *knowing* crime-relevant information is equivalent to a brain state. But it is not, knowing something is a behavioral or person-level state of affairs:

> This characterization depends on a confused conception of knowledge. Neither knowing something nor what is known – a detail about a crime, for example – is stored in the brain . . . Suppose, for example, a defendant has brain activity that is purported to be knowledge of a particular fact about a crime. But, suppose further, this defendant sincerely could not engage in any behavior that that would count as manifestation of knowledge. On what basis could one claim and prove that the defendant truly had knowledge of this fact? We suggest that there is none; rather, as with a discrepancy regarding lies and deception, the defendant's failure to satisfy any criteria for knowing would override claims that depend on the neuroscientific evidence.[40]

But in focusing on this example, Pardo and Patterson are constructing something of a straw man. As noted previously, Farwell is not necessarily representative of the best scientific work being done on the interpretation of the P300 CIT. Indeed, Farwell's characterization of the brain as something that simply records and stores knowledge has been criticized by leading proponents of the P300 CIT. For example, J. Peter Rosenfeld wrote a lengthy critique of Farwell's interpretation of the test back in 2005. In it, he identified the problems with Farwell's analogy between the

39 Farwell, L. and Smith S. S. 'Using Brain MERMER Testing to Detect Knowledge Despite Efforts to Conceal' (2001) 46 *Journal of Forensic Science* 135–43.

40 Patterson and Pardo (2013), 101–2.

brain and a video recorder and noted that the best available neuroscientific and psychological evidence indicated that the act of remembering or recollecting information is highly fallible and reconstructive, involving a complex interplay between conscious and subconscious reasoning processes. Of course, Farwell is not strictly speaking a straw man: he really exists, and he really has pushed for the use of this technology in the courtroom. So I am not claiming that there is no danger here or that Pardo and Patterson are completely wrong to warn us about it. My only point is that Farwell is not particularly representative of the work being done in this field, and that there are others that are live to the dangers of assuming that the P300 signal does anything more than provide inductive evidence of knowledge. As far as I am aware, no one else makes the mistaken assumption that the detection of the P300 is criterial evidence of knowledge. They all stress that further inferential steps are needed to reach such a conclusion. This is why the original term for the test (the guilty knowledge test) has been abandoned in favor of 'concealed information' or 'recognition detection'. When the test is framed and understood in this light, it is much less obvious how the mereological fallacy might arise.

Pardo and Patterson are also claiming that there is an important distinction between a person *knowing* something and the neurological states of that person. Knowledge is a state pertaining the whole, whereas neurological states are subparts of that whole. This seems fair enough. But I and others do not conceive of the test in these terms. The usual conception and understanding is that the test determines whether a *person's brain* (as opposed to the person themselves) recognizes information. A person's brain could signal recognition of a stimulus without the person themselves recognizing the stimulus. Why? Because we know that large parts of what the brain does are subconscious (or subpersonal – if we assume the personal is defined by continuing streams of consciousness). Nevertheless, figuring out whether a subject's brain recognizes a stimulus is still forensically useful and need not be confused with assuming that the person recognizes the stimulus. Knowing that a person's brain recognizes information can help to place their physical body at the scene of a crime (provided other potential exposures to the information can be ruled out).[41] This is surely forensically useful, even if it turns out

[41] This is an important point. All proponents of the P300 that I am aware of acknowledge that the test's utility is ruled out if the information being presented to the subject could have been presented to them in a context other than that of the historical incident under

that other evidence suggests that the person (as opposed to their brain) does not recognize the information and lacks 'guilty' knowledge. As long as we are clear about this distinction, we have little to fear from the mereological fallacy.

The Third Critique: The Status Quo Fallacy

This brings me to a final criticism of Pardo and Patterson's argument. This is probably the most important. A major problem I have with Pardo and Patterson's discussion of brain-based lie detection is how isolated it feels. They highlight the empirical and conceptual problems with this form of evidence without considering the significance of this evidence in its appropriate context. I am happy to grant that the P300 CIT has empirical limitations, and that there is a slight risk that proponents of the P300 CIT will commit the mereological fallacy (though I think this can be easily addressed). But I would insist that the significance of these limitations and risks cannot be assessed in isolation. It can only be assessed in the relevant context.

By relevant context here I mean the existing *evidential status quo* with respect to deception detection or lie detection. It well known that humans have a bias toward the existing status quo, and as Bostrom and Ord[42] point out, this bias often obscures or blinds them to ways in which new technologies or interventions can improve things relative to that status quo. There is, it seems, a terrible tendency for people to focus on the costs and benefits of new technologies without focusing on the costs and benefits of the existing situation. The result of this is that they over-emphasize the costs of the intervention and ignore the costs of the current status quo.

This 'status quo' fallacy must be factored in as a counterveiling consideration that points in the opposite direction to the 'mereological fallacy'. In other words, when assessing the strengths and weaknesses (both empirical and conceptual) of these brain-based tests, we also need to assess the strengths and weakness of the current methods we use

investigation. The risk of this undermining the utility of the test can, however, be mitigated by ensuring that they are tested on multiple pieces of information and by ensuring that investigation protocols are carefully established to minimize the risk of accidental exposure.

42 Bostrom, N. and Ord, T. 'The Reversal Test: Eliminating Status Quo Bias in Applied Ethical Reasoning' (2006) 116 *Ethics* 656–79.

for determining whether a witness is lying or whether a suspect was at a crime scene. Although some of these methods are empirically and conceptually robust (e.g., DNA testing), they are not always available, and other methods leave much to be desired on both an empirical and conceptual front. For example, it is extremely common to use methods like robust questioning, cross-examination and interrogation to work out whether someone participated in a crime or is being deceptive. The people engaging in these forms of questioning often use the verbal or behavioral responses from the interviewee to make inferences about what someone knows or does not know. But the theory underlying these methods is poorly worked out, they are not particularly empirically reliable, and their error rates are largely unknown.[43] So even if behavioral criteria determine what it means for a subject to know something, there are still all sorts of behavioral signals that are misinterpreted and that can mislead us. Is someone hiding something if they are being fidgety? Or if they look nervous and blink too often? What if they change their story? We routinely make inferences from these behavioral signals without thinking about their epistemic value.

This matters. One of the points that I and others have been making in relation to the P300 CIT is that it provides a neurological signal, from which we can make certain inferences, and that it comes with known error rates and precise protocols for its administration.[44] In this respect it seems to have a *comparative advantage* over many of the other methods we use for making similar inferences. And this is why we should take it seriously. Even if it does carry with it the risk that legal actors will commit the mereological fallacy, that risk has to be weighed against the risk that those actors will commit the status quo fallacy and are confused or overly confident about existing, evidential methods. If the risks of error attached to the latter outweigh the risks of error attached to the former, Pardo and Patterson's argument might be a good deal less persuasive. And I think the latter risks do outweigh the former, particularly in certain contexts. One such context is pretrial plea-bargaining in criminal cases. I have made the case for using the P300 CIT in this

[43] See Meixner (2012); Schauer, F. 'Can Bad Science Be Good Evidence? Neuroscience, Lie Detection and Beyond' (2010) 95 *Cornell Law Review* 1191–219; and Schauer, F. 'Lie Detection, Neuroscience, and the Law of Evidence (2012) *Virginia Public Law and Legal Theory Research Paper* 2012–49.

[44] See Schauer (2012 and 2015), ibid.; Meixner (2012), ibid. and (2015); and Danaher (2015a).

context at greater length elsewhere (2015a) but I will offer a brief summary here to round out this critique.

Plea-bargaining is the practice whereby someone pleads guilty to a lesser charge to avoid the risks associated with being tried for a more serious offence. It is widely known that (in all but a few exceptional cases) criminal defendants have less bargaining power than prosecutorial authorities. As a result, prosecutors are often able to drive reasonably hard plea bargains. So much so, in fact, that there is a problem with innocent defendants pleading guilty to offences they never committed.[45] It seems that some people are so worried about the risks associated with being tried for the more serious offence, and so incapable of proving their own innocence, that they are willing to take the hit. Studies of exonerations suggest that up to 10 per cent of those who plead guilty later turn out to be innocent.[46]

There are many possible causes of the innocence problem. Long-standing structural and political issues are part of the problem, as are the idiosyncrasies of particular cases and personalities. But at the heart of them all is a basic *signaling problem*. Innocent defendants feel the pull of the guilty plea because they frequently have no way in which to credibly signal their innocence to the prosecutors. Covey[47] has argued that this signaling problem can be solved by introducing a 'subwager' into the pretrial bargaining game that is played between prosecutors and defendants. The subwager is akin to a bet that a truly innocent defendant would be willing to take, while a guilty one would not. In other words, it is a bet with asymmetrical risks: it is high risk to the guilty defendant but low risk to the innocent. Hence, only the truly innocent defendant is likely to take the bet. I argue that voluntary submission to a P300 CIT would constitute taking such a bet. Why? Well, the P300 CIT has an asymmetrical risk profile. It is more likely to falsely exonerate than to falsely inculpate.[48] This means it is relatively low risk to the innocent defendant: they are unlikely to falsely inculpate themselves by submitting to the test. But it is also relatively high risk for the guilty defendant. You might wonder how this could be if it is biased in favor of false negatives.

45 See Rakoff, J. 'Why Innocent People Plead Guilty' *New York Review of Books* 20 November 2014 and Covey, R. 'Signaling and Plea Bargaining's Innocence Problem' (2009) 66 *Washington and Lee Law Review* 73–130.

46 Rakoff (2014), ibid. 47 Covey (2009).

48 I use these terms with some caution. As outlined in the previous section, the test by itself does not exonerate or inculpate, it merely provides evidence that can be used to infer those possibilities.

The answer is that this biasing means that the strength of the inductive inference that can be drawn from a positive result is greater than it would be from a test without this feature. Thus, a guilty defendant runs a serious risk by voluntarily submitting to the test. On top of this, the P300 CIT provides a better way to signal innocence than other mooted subwagers such as Covey's own proposal of voluntary submission to robust interrogation. Consequently, the P300 has significant advantages over existing methods for extracting evidence in this particular context.

There is much more to be said, of course.[49] If the P300 CIT is to be used in this way, various safeguards will need to be put in place to ensure that it is not abused and that it is administered with due care, skill and diligence. Nevertheless, we see here the potential for this technology. It can improve things relative to the current status quo. This gives us reason for optimism rather than skepticism when it comes to the future of brain-based lie detection.

Conclusion

To briefly sum up, Pardo and Patterson offer an interesting and philosophically sophisticated critique of brain-based lie detection. They highlight its empirical weaknesses and suggest that one of the primary dangers with this technology is that the legal actors who make use of it will be prone to commit the neurolaw mereological fallacy. Though I agree that this is a fallacy, I argue that it is not that dangerous, at least in the case of evidence drawn from the P300 CIT. This is for three reasons. First, I think the empirical strengths of this technology are greater than Pardo and Patterson let on. Second, I think the risk of actual proponents of the technology committing the mereological fallacy is slight and easily addressed. And third, and most important, the risk of committing the mereological fallacy must be weighed against the status quo fallacy, i.e., the risks associated with making faulty inferences from other types of evidence. It is suggested that the latter can be higher than the former and that this blunts Pardo and Patterson's skepticism and provides reason for optimism.

[49] Danaher (2015a).

8

Is Brain Reading Mind Reading?

PIM HASELAGER AND GIULIO MECACCI*

Introduction

Recent advances in cognitive neuroscience have made it possible to decode or interpret mental states of human beings by measuring brain activity. This is often referred to as 'brain reading': using brain measurements for mind reading. The concept of 'brain reading' is complex. In this chapter, we will examine aspects of its meaning, and at least some of the issues that need to be addressed in order to assess any claim that brain reading has been achieved. We will attempt to do so by discussing the concepts of brain reading, mind reading, and reading. We will indicate some consequences of our analysis by addressing its implications for the Fifth Amendment to the US Constitution. We will argue that, in its current stage, brain reading should not yet be considered to instantiate a case of mind reading. By this we mean that it does not decode the combinatorial syntax and semantics of mental states. Following Pardo and Patterson's[1] analysis of the Fifth Amendment, we suggest that, for this very reason, the Fifth amendment does not apply to brain reading yet. We leave open the possibility that it may so apply when the challenges of syntax and semantics have been met.

Reading the Brain to Read the Mind?

In the *Syracuse Herald* of 1919, a drawing of a 'thought recorder' was published,[2] and the accompanying text says: 'The thought recorder is an instrument recording thoughts directly by electrical means, on a moving

* The authors would like to thank Lorenzo Palamenghi for valuable comments on earlier drafts.

1 Patterson, D., and Pardo, M., *Minds, Brains and Law* (Oxford: Oxford University Press, 2013).

2 Anon, 'This Machine Records All Your Thoughts' (1919) *The Syracuse Herald* Available at: www.paleo-future.blogspot.nl/2007/05/this-machine-records-all-your-thoughts.html.

paper tape'. Shown is a headband not unlike the currently commercially available dry electrode wearable EEG headband of, e.g., Emotiv.[3] Then, in theory (as now, in practice), the aim was clear: to measure aspects of neuronal activity in order to access the content and type of mental states and processes[4]. Various techniques exist for the measurement of brain processes. Functional magnetic resonance imaging (fMRI) measures blood flow via the magnetic properties of hemoglobin that carries oxygen to brain areas that have been activated in order to restore energy. Electroencephalography (EEG) measures the collective spikes of billions of neurons. Single-cell recordings measure the activity (discharges) of individual cells. Each technique has its own advantages and disadvantages, but all of them involve technologies directed at measuring specific features of the brain. What makes brain measurements candidates for the label 'brain reading' is that the measurements are taken in order to 'read someone's mind'. These measurements could be put to other uses; in Brain-Computer Interfacing, they might be used to drive an application (e.g., a cursor on a computer screen),[5] for example, These possibilities are not relevant for present purposes because although they do *use* brain measurements, there is no attempt to decode the measurements or to interpret their mental meaning.

'Mind reading' in itself is nothing new. For as long as there has been social interaction, the agents involved have been trying to 'mind read': to guess or infer the intentions, beliefs, desires, feelings or capacities of others. More specifically, one may think of the following examples of some traditional psychological categories: beliefs (I think that unicorns exist), desires (I want a glass of beer), knowledge (a triangle is a polygon with three edges and three vertices), intention (I'll climb that mountain), occurrent thought (did I see this before?), dispositions (character traits, personality features: Mary is a social, extrovert, hot-tempered person), qualia (seeing red, tasting wine, feeling happy), etc. By observing the behaviour of others (which may, but need not, include linguistic utterances), we can attribute a wide variety of mental states and processes to them. For instance, on the basis of seeing brisk, sudden movements of

[3] www.emotiv.com.

[4] The difference between content and type of a mental state can be illustrated as follows: one can hope or fear that 'the man with the white hair wins the election'. Such mental states would have the same content, but express a different attitude toward it. Of course someone could also hope that 'the man with the *red* hair wins the election'. Here the attitude is the same (hope), but the content different.

[5] Van Gerven, M. et al., 'The Brain-Computer Interface Cycle' (2009) 6 *Journal of Neural Engineering*, 41001.

an individual, accompanied by loudly voiced obscene language, we may infer that the person is angry, and, from further contextual information (e.g., frantically searching his pockets) that the person believes he lost something (or has been robbed).

What brain reading adds to the traditional interpretation of the behaviour of others is a different *mode of observation*. That is, another way of inferring the content and nature of mental states has become available, i.e., by observation (measurement) of brain activity. Through this new method, it may become possible to see new or more refined distinctions between mental states and processes, (though this need not necessarily follow). And, brain reading opens up the opportunity to collect information about mental states – that may or may not be actively concealed from external view – without recurring to directly observable behaviour.[6] Of course, together with the novel possibilities it portends, this additional mode of observation (measuring brain activity) may bring about new and/or profound consequences, for instance, regarding privacy and law.

The Meaning of Reading

Reading is a rather complex process of constructing meaning from a written text. It consists of various subskills (e.g., discriminating letters, recognizing words), requires practice, and demands sensitivity to more than one interpretation, depending on context. More extensively, it can be described as the decoding of encoded information, by recognizing and categorizing representations (usually words or word combinations), extracting their meaning (sense and reference), and parsing syntax (based on word order) in order to grasp the grammatical structure. In relation to representation, it is customary[7] to make a distinction between the physical aspects of it (the 'vehicle', e.g., specific spiking frequencies, or deoxygenated blood flow in the brain) and the meaning that the vehicle

[6] To a minor extent, this was already possible with classic investigation methods like the polygraph, which aims at picking up physiological signs. More information on the relationship between different mind reading technologies can be found in G. Mecacci & P. Haselager 2017, Identifying Criteria for the Evaluation of the Implications of Brain Reading for Mental Privacy, Science and Engineering Ethics. https://link.springer.com/content/pdf/10.1007/s11948-017-0003-3.pdf.

[7] See Cummins, R., *Meaning and Mental Representation*, (Cambridge, MA: MIT Press, 1989); Dretske, F., *Explaining Behavior: Reasons in a World of Causes*, (Cambridge, MA: MIT Press, 1998); Uithol, S. et al., 'What Do Mirror Neurons Mirror?' (2011) *Philosophical Psychology*, 24(5), 607–23.

carries (its content). Additional aspects of a representation are the user (the reader) and the objects or events that the content may refer to or be about.

Within cognitive neuroscience, brain reading comes down to measuring various features of brain activity (classifying or identifying the vehicles) and correlating them to cognitive processes or states or dispositions. This goes via a sequence of stages. First, brain activity is measured and recorded while a person engages in a specific cognitive task, e.g., entertains a particular thought. Second, algorithms learn which brain processes correlate with these cognitive tasks, so they are trained to learn a particular mapping between cognition and brain activity. In a third step (the 'test phase') these trained algorithms are applied to new brain measurements of persons, and then infer ('predict') which cognitive processes these persons are most likely to be engaged in. Once the correlation has been established, one can, at least in principle, infer with a degree of certainty from observed brain activity that a cognitive state or process has taken, is taking or will take place.

For the purpose of this chapter we will focus on beliefs and desires involving propositional attitudes.[8] Brain reading has been applied with varying degrees of success to reading an individual's intentions,[9] the willing of voluntary actions,[10] the content of visual perceptions,[11] memories,[12] thoughts,[13] dreams[14] and emotions,[15] the nature of political

[8] Fodor, J., 'Propositional Attitudes' (1978) *The Monist*, 61(4), 501–23.

[9] Soon, C. S. et al., 'Predicting Free Choices for Abstract Intentions' (2013) *Proceedings of the National Academy of Sciences*, 110(15), 6217–22.

[10] Soon, C. S. Brass, M. Heinze H. J. and Haynes, J. D. 'Unconscious Determinants of Free Decisions in the Human Brain' (2008) *Nature Neuroscience*, 11, 543–5; Filevich, E. et al., 'Brain Correlates of Subjective Freedom of Choice' (2013) *Consciousness and Cognition*, 22(4), 1271–84.

[11] Kay, K. N. et al., 'Identifying Natural Images from Human Brain Activity (2008) *Nature*, 452(7185), 352–5; Schoenmakers, S. et al., 'Linear Reconstruction of Perceived Images from Human Brain Activity' (2013) *NeuroImage*, 83, 951–61; Stansbury, D. E., Naselaris, T. and Gallant, J. L., 'Natural Scene Statistics Account for the Representation of Scene Categories in Human Visual Cortex' (2013) *Neuron*, 79(5), 1025–34.

[12] Peth, J. et al., 'Memory Detection Using fMRI – Does the Encoding Context Matter?' (2015) *NeuroImage*, 113, 164–74.

[13] Huth, A. G. et al., 'Natural Speech Reveals the Semantic Maps That Tile Human Cerebral Cortex' (2016) *Nature*, 532, 453–8.

[14] Horikawa, T. et al., 'Neural Decoding of Visual Imagery during Sleep' (2013) *Science* (New York, NY), 3, 639–42.

[15] Huis in 't Veld, E. M. J., and de Gelder, B., 'From Personal Fear to Mass Panic: The Neurological Basis of Crowd Perception' (2015) *Human Brain Mapping*, 36, 2338–51.

preferences,[16] sexual orientation,[17] the levels of pain,[18] experienced pleasantness of wine tasting,[19] and consciousness.[20] In the *Oxford Handbook of Neuroethics* one of the leading experts in the field summarizes the state of the art in the following way:

> By combining fMRI with pattern recognition, the field of 'brain reading' has made huge progress in the last few years. It has been possible to read out very detailed contents of a person's thoughts, including detailed visual percepts and ideas, memories, and even intentions and emotions. It is possible to read out implicit and even unconscious mental states, such as unconscious percepts and decisions.[21]

It is hardly surprising that this research has led to rather strong claims and expectations (hopes as well as fears) in society, as expressed in popular media.[22] Different assessments, ranging from extreme skepticism to outright science fiction, about current possibilities and future potential, have tended to obscure philosophical and societal analyses and debates. In practice brain reading encounters substantial challenges, and most applications are currently far from being sufficiently reliable or practically useful (a framework for the assessment of brain

[16] Lamme V. A. F. 'What Introspection Has to Offer, and Where Its Limits Lie' (2010) *Cognitive Neuroscience*, 1(3), 232–40.

[17] Poeppl, T. B. et al., 'Connectivity and Functional Profiling of Abnormal Brain Structures in Pedophilia' (2015) *Human Brain Mapping*, 36(6), 2374–86.

[18] Cowen, R. et al., 'Assessing Pain Objectively: The Use of Physiological Markers' (2015) *Anaesthesia*, 70(7), 828–47.

[19] Plassmann, H. et al., 'Marketing Actions Can Modulate Neural Representations of Experienced Pleasantness' (2008) *Proceedings of the National Academy of Sciences of the United States of America*, 105(3), 1050–4.

[20] Blume, C. et al., 'Across the Consciousness Continuum – from Unresponsive Wakefulness to Sleep' (2015) *Frontiers in Human Neuroscience*, 9, 1–14.

[21] Haynes, J. D., 'Brain Reading: Decoding Mental States from Brain Activity in Humans' in J. Illes and B. J. Sahakian, eds. *Oxford Handbook of Neuroethics* (Oxford: Oxford University Press, 2012), 1–8.

[22] ten Broeke, A., 'De Scanner Zegt het. Dus het is Echt. Punt uit' (The Scanner Says So. Hence It Is Real. Period') (2014) *De Volkskrant*, www.volkskrant.nl/opinie/de-scanner-zegt-het-dus-het-is-echt-punt-uit~a3765206/; ten Broeke, A., 'U Bent Pedofiel, Zegt de Scanner' ('You Are a Paedophile, the Scanner Says)' (2011) *Trouw*, available at: www.trouw.nl/tr/nl/6700/Wetenschap/article/detail/3097538/2011/12/29/U-bent-pedofiel-zegt-de-scanner.dhtml Sample 2012; Wolpe, P. R., 'Is My Mind Mine?' (2009) *Forbes*, available at: www.forbes.com/2009/10/09/neuroimaging-neuroscience-mind-reading-opinions-contributors-paul-root-wolpe.html; Roth, M., 'CMU Knows What's on Your Mind' (2009) *Pittsburgh Post-Gazette*, available at: www.post-gazette.com/science/2009/01/04/CMU-knows-what-s-on-your-mind/stories/200901040263.

reading applicability can be found in recent work by Mecacci and Haselager.[23] A foundational challenge relates to the nature of brain reading. Is it reading at all?

Is It reading?

A number of questions can be raised about the extent to which the concept of 'reading' applies to the process of brain reading. Here, we will focus on two aspects in particular, syntax and semantics, starting with the latter.

Semantics

In relation to semantics, one of the crucial questions is the extent to which brain measurements can currently give access to mental *content*. To put it another way, we might legitimately wonder whether brain measurements enable the interpretation of what someone's thought is about, or the experiential nature of someone's feeling. Can it really be said, for example, that currently available processes provide insights into a person's mental states, similar to the insights gleaned by directly asking the person to describe these states? Might it be that the results of neuroscience are more akin to an educated guess about what might be going on in somebody's mind at quite a general and abstract level?

Certainly, using the representational terminology introduced earlier, there is a difference between understanding the meaning of a brain state (accessing the content) and simply categorizing the vehicles that carry it. These process can be understood as different points on the same continuum: categorizing vehicles is one of the first steps toward fully understanding the meaning they carry. An example to illustrate this continuum could be the following: consider a person that is exposed to a system of signs (vehicles) of a language she does not understand (in our case, for instance, Japanese). In the beginning, that person can only distinguish them from one another, on the basis of their shape. She has no idea, though, about their meaning. Give that person a pocket vocabulary, and she will start being able to categorize the signs she can recognize, connecting them to some meaning. Due to the limitations of her short manual, her understanding will be limited, superficial and prone to mistakes. Several years later, after spending some time in Japan and

[23] Mecacci and Haselager, (2017). (above at n.6).

studying the language extensively, she might have developed a rich and usable understanding of the Japanese language.

By analogy, current brain reading is moving from the stage of distinguishing between vehicles to the stage our language learner was at when she acquired her first vocabulary. The pocket vocabulary for the brain reader is still small and oftentimes very general; only relatively high-level semantic categories can be distinguished. In short, brain reading methods can currently recognize – and discriminate among – a rather limited number of vehicles (brain states), and the meaning (mental states) those vehicles can be connected to, is often abstract and lacking detail. It is therefore not surprising to find the author quoted earlier softening his claim almost to the point of retraction:

> Currently, it is not possible to directly 'read' the 'language' of the brain, that is, to identify mental states based on a systematic interpretation of the corresponding brain states.[24]

Our analysis is consistent with this latter claim and in direct opposition to his earlier one.

Syntax

In relation to syntax, the question is whether or to what extent brain reading techniques enable us to access the constituent structure[25] of concatenations of symbols. That is: can an algorithm be trained such that a reliable distinction can be made between an unstructured collection of representations and a proper syntactical structural relation between representations? As the standard example goes: there is a difference between 'There is something about John, Mary, and love', and 'John loves Mary'. The first case (the unstructured collection of representations) does not provide information about who loves and who is loved, but the second case does. Similarly, there is a difference between 'John loves Mary' and 'Mary loves John'. Combinatorial syntax enables to distinguish between the latter two sentences and hence to distinguish their (potentially dramatically) different meanings.

[24] Haynes (2012) at p. 3.

[25] Fodor, J. A. and Pylyshyn, Z. W., 'Connectionism and Cognitive Architecture: A Critical Analysis' (1988) *Cognition*, 28(1–2), 3–71; and Fodor, J. A., 'The Language of Thought' (1975) *The Philosophical Review*, 20(1), 59–63.

Huth et al. investigated how the meaning of language is represented in the brain. In principle, claim the authors, this 'suggests that the contents of thought, or internal speech, might be decoded'.[26] The scientists identified – with variable margins of error – different cortical areas that encode for semantic mental states belonging to 12 semantic domains, such as, e.g., 'person', 'visual' or 'time'. Although these results constitute a remarkable advance in neuroscientific research, the degree of abstractness and the number of states the scientists were able to discriminate with reasonable reliability are far removed from what we could commonly intend by the concept of 'reading', where a reader is fully competent on a certain language, and capable of understanding the meaning of a large corpus of sentences. Even if it is possible to access to some extent the meanings of individual representations (but see the point about meaning earlier), there is no evidence that the further question of addressing syntax, i.e., the structural relations between semantic elements, can be captured via current brain reading techniques. Hence, we have to disagree with Huth et al.'s claim that the contents of internal speech can be grasped, until it has been shown that not only representations belonging to different semantic categories can be read out, but also the combinatorial syntax combining them into meaningful statements (sentences or propositions).

Implications of the Current Limitations of Brain Reading

Pardo and Patterson[27] discuss how neuroscientific evidence could potentially be used against criminal defendants and how constitutional provisions can protect defendants by limiting government evidence gathering. Specifically, they discuss the Fourth and Fifth Amendments to the US Constitution that, respectively, prohibit unreasonable searches and seizures and provide privilege against self-incrimination. We would like to focus on their discussion of the Fifth Amendment to investigate what the consequences are of our analysis of brain reading.

Regarding the Fifth Amendment, Pardo and Patterson say[28] that it offers protection for defendants in case of attempts to procure evidence that meets three formal requirements: (1) compelled, (2) incriminating, and (3) testimonial communications. Compulsion would be the case when a suspect, for instance, is forced physically (or through threats) to

[26] See Huth et al. (2016). [27] Patterson and Pardo (2013); Chapter 6. [28] Ibid., 162.

undergo fMRI or EEG scanning. Incrimination is the case when the obtained information will be used in a criminal prosecution of the defendant, e.g., when the results of brain reading would be used as evidence that the defendant committed a crime. The third requirement (testimonial) is the most interesting, but also quite complex. We give only the upshot of Pardo and Patterson's analysis here. In their view, basically the Fifth Amendment does not protect the physical instantiations of a 'testimony' (e.g., a document or a recording of a statement in itself), but what gets expressed in them (i.e., the mental *content* that finds its physical instantiation through a document or recording). In other words, it is the content of a mind[29] that makes a document or recording 'testimonial'. It is because of this that the Fifth Amendment offers protection against unwarranted access to documents or recordings reflecting mental content because accessing them would constitute self-incrimination. As they say:

> the scope of the privilege turns on whether the government is making incriminating use of the contents of the defendants' mental states . . . In focusing on the content of mental states, the privilege appears to focus on what philosophers refer to as 'propositional attitudes' . . . The privilege appears to apply to the content of these propositions . . . and limits the government from making compelled, incriminating use of them.[30]

Given our earlier analysis, the emphasis on access to mental content has an interesting consequence for the scope of the Fifth Amendment regarding brain reading. To the extent that current brain reading is still mainly coming down to classifying vehicles and only limitedly capable of accessing (decoding, interpreting) the content of mental states, the Fifth Amendment does not seem to offer protection. In other words, in the continuum that goes from mere discrimination of signs, to full understanding of meaning (content), brain reading methods are arguably still very close to the former side. Insofar as this is the case, the safeguarding of mental content by the Fifth Amendment, might just not apply.

However, when brain reading technology would provide a substantially more detailed access to the content of mental states, by including more semantic domains at much finer levels of detail, and managing to capture syntax (i.e., capture the structural relations between representations to combine them into an overall proposition), then the data provided would start to fall under the Fifth Amendment tutelage. In short,

[29] Ibid., 165. [30] Ibid., 165–6.

current brain reading technology may fall outside the scope of the privilege that the Fifth Amendment affords defendants precisely because of its current shortcomings, and only as long as these last.

Conclusion

Brain reading technology is developing fast, but despite certain claims about success, it is doubtful whether current applications actually amount to the 'reading' of mental content in a full sense of the word. Rather, today's brain reading technology appears to engage in a kind of 'vehicle' classification that barely provides with loose and highly abstract interpretations of mental content. Legally, this may imply that information obtained through current brain reading technology falls outside the scope of the Fifth Amendment, at least as analyzed by Pardo and Patterson. That said however, progress in brain reading is very fast. It is therefore recommendable to consider the aspects of syntax and semantics, discussed earlier, as challenges for brain reading rather than as definite obstacles. The applicability of legal safeguards regarding unwarranted access to mental content is in need of constant reevaluation. The safeguards themselves may need significant revision in light of neuroscientific progress to remain societally relevant. Given the enormous potential consequences of brain reading, it is crucial to keep track of its practical challenges and progress.

PART III

Implications for Courts and Defendants

9

Unlucky, Bad, and the Space in Between: Why Criminologists Should Think More about Responsibility

PETER RAYNOR

The simple things you see are all complicated[1]

In the early 1970s I was working as a probation officer, supervising offenders in the community and providing information about defendants to magistrates and judges in order, supposedly, to help them to make appropriate sentencing decisions. In the course of this work, to cut a long story short, I met a few people who committed crimes in a very deliberate and planned way, for personal gain or advantage, in circumstances where they clearly had every opportunity to avoid crime and to meet their needs or satisfy their desires in legitimate ways. Much more often, however, I was dealing with people who did not seem to be showing such a high degree of control and self-direction in their offending. Some were experiencing extreme poverty; some were the products of disorganised families, where there was little opportunity for normal social learning; some were offending while their self-control was impaired through alcohol or drugs; some suffered a range of physical or mental illnesses (though not usually of a kind that *compelled* them to offend); some had difficulty controlling the impulse to throw a punch or place yet another bet. Many had difficulty seeing beyond the short-term situation in which they offended, while others lacked understanding of how their behaviour affected others, particularly their victims. Some were victims themselves, living with the consequences of early abuse, while others seemed trapped into lifestyles in which offending was normal, and previous offending had closed off most legitimate opportunities. This brief list does not exhaust the commonly encountered antecedents of offending: criminologists know about, and workers in criminal justice encounter, a wide range of

[1] The Who, 1966.

crimes, of people who get themselves into trouble and of accounts (both professional and lay) that seek to explain how crime occurs. Most of the time, however, I seemed to be dealing with people who were clearly not compelled to commit offences, but who also found it very difficult to avoid doing so and were often under circumstantial pressures that made offending appear, at least in the short term, to be a viable choice.

I also met, less frequently, people whose behaviour clearly resulted from mental illness and was beyond their control. However, between these cases and the other rare situations in which offences had been committed without any obvious pressures at all, simply because somebody wanted something they could not legitimately have, were most of the people with whom I worked. In other words, most people who were offending seemed to be doing so in situations when it would be possible but difficult to avoid doing so, given their circumstances, their limited material and personal resources, and their established habits of thinking and behaviour. Occasionally, if a homicide was involved, the law might recognise extreme situational pressures or mental stresses by deploying the concept of diminished responsibility, but this approach was not available for routine everyday offences. In between the extremes of, on one side, freely chosen selfish antisocial crimes and, on the other side, actions that genuinely could not be avoided (in other words, between bad and unlucky) lay a broad territory that many offenders inhabited most of the time. For them, offending was a choice, but an unwise one, in circumstances that they usually had not chosen and that offered few good alternatives. Many offenders regretted their offending and hoped to avoid it in future, when given the opportunity to reflect on it in calmer circumstances. How far, I wondered, should we blame such people for making these bad choices? Or was all offending simply the product of forces beyond the offenders' own control?

Crimes and Diseases

While I was still involved in practical work with people who were in trouble with the law, the philosopher Antony Flew published his short book *Crime or Disease?*[2] which set out to address some of the same questions. His concern was to refute the arguments of deterministic psychiatrists and psychologists who held that because all behaviour was

[2] Flew, A. *Crime or Disease?* (London: Macmillan, 1973).

causally determined, none of it could be regarded as freely chosen. Responsibility, guilt, blame, desert were all prescientific concepts and illusory. More recently, because some changes in the brain that are relatively reliable precursors of behaviour can now be observed and measured, the standard-bearer for such claims has been neuroscience (for example, 'Free will, as we normally understand it, is an illusion' write Greene and Cohen[3]), but well before the 1970s it had been observed that brain injuries and brain surgery could alter behaviour, moods and even beliefs. In addition, electrical stimulation of some parts of the brain could sometimes produce movements that the subject believed to be voluntary. Flew was, to say the least, unconvinced by claims that psychological determinism could abolish choice and responsibility, and his arguments bear revisiting in the context of more recent controversies.

Diseases, argues Flew, and in particular mental diseases, are distinguished by the fact that they create incapacities and reduce our ability to act as we choose. In the absence of disease (which, to avoid circularity, needs to be evidenced by something other than the behaviour we are trying to explain) the capacity for choice exists, based on the capacity to exercise free will. Our experience of free will is based on a 'familiar difference' between what Flew calls 'moving' (i.e., voluntary action) and 'motion', which is involuntary movement such as a knee-jerk reflex or a sneeze. The difference is illustrated by the contrast between 'I moved my arm' and 'my arm moved (although I didn't move it)'. Flew argues that such 'familiar differences' are undeniable: 'no discovery ... could ever abolish familiar differences, or invalidate distinctions founded solely upon such differences.'[4] In addition he points out, surely correctly, that the commonsense and ordinary language meanings of phrases like 'I did it of my own free will' are not metaphysical claims to be an uncaused cause, but are simply accounts of agency in relation to movings (e.g., 'I moved my arm of my own free will') or statements about the absence of coercion (e.g., 'I did it of my own free will, nobody forced me'). He argues that even if full determinism exists (and it is not clear to me what would count as conclusive evidence for this), choices can be causes and have consequences without themselves being uncaused. I would add that reasons for action can be explanatory ('this is how I came to do this')

[3] Greene, J. and Cohen, J. 'For Law, Neuroscience Changes Nothing and Everything' in Zeki, S. and Goodenough, O. eds. *Law and the Brain* (Oxford: Oxford University Press, 2006), 207–226.

[4] Flew (1973), 103.

or justifying ('this is why I thought this was the right thing to do') and that these can produce different reasons without invalidating each other.

These arguments made fascinating reading for a young probation officer, dealing every day with issues arising from people's imperfect control over their own actions or their poor choices about them. In retrospect it also seems clear that more recent deterministic claims arising from neuroscience are not logically different from those made by some psychiatrists and psychologists in the 1970s, and Flew's arguments still work. However, these arguments applied most clearly to fully free actions (moving) and unfree behaviour (motion). Some thought was still needed about what they meant for the broad range of intermediate cases of choices made under pressures that fall short of coercion and impulses potentially subject to control that were not controlled. What did they imply about the culpability of people whose offending seems to be as much a natural consequence of their circumstances as a deliberate choice? These concerns can be illustrated by considering some of the everyday difficulties of interpreting and allocating responsibility and blameworthiness in criminal justice.

Social and Psychological Information in Sentencing

A hundred students are listening to a lecture on punishment and rehabilitation. The lecturer introduces a sentencing exercise, based on cases in his past experience: a male burglar, 19 years old, with two previous convictions for theft that resulted in fines, has been convicted of breaking into a house while the owners were away, stealing a computer, TV and jewellery worth £1000. He was arrested trying to sell the jewellery. What sentence would the students give him? The vast majority opt for an immediate custodial sentence. Then they are given some more information: his parents separated when he was four years old; he was brought up in local authority care until he was 16; he had a job in a shop but lost it when the shop went bankrupt. He has no educational qualifications and poor literacy and is technically homeless so he receives limited welfare payments. Asked to vote again, far fewer students choose the immediate custodial sentence. A few more facts are added: he was drunk when he did the burglary, he drinks every day with his friends, half his income (when he has one) goes on drink and drugs, and he has not finished paying his outstanding fines. The majority choice now becomes a period of supervision by the probation service combined with assistance to reduce his use of alcohol and drugs. The social information appears to have had two effects: not only has it provided information about needs that can be addressed through appropriate sentencing to reduce the probability of further offences, but it has also reduced the severity of the proposed sentence. It has affected the students' views of how much punishment the offender deserves and of how blameworthy he is for the current offence.

Something like this often seemed to be the impact of the hundreds of social enquiry reports (now called presentence reports) that I prepared about the background, behaviour and personality of defendants appearing in court.[5] In many cases it seemed that the background information was influencing perceptions of an offender's blameworthiness as well as the sentencer's thinking about what results a sentence might be able to achieve. Adding information derived from neuroscience to the available knowledge about an offender might assist understanding of internal as well as external constraints (for example, a history of brain injury or concussive accident might make it more difficult for some people to control their tempers or to resist other kinds of impulses), but it does not appear that this kind of information has a different logical force from the kinds of social and personal information already mentioned. In particular, it does not abolish responsibility unless it demonstrably amounts to compulsion. Short of compulsion, such information might affect judgement about degrees of blameworthiness, but responsibility remains. Pardo and Patterson have convincingly argued against claims that knowledge derived from neuroscience invalidates legal views of responsibility.[6]

Theories of punishment distinguish between the retributive rationale, which relates the amount of punishment to what the seriousness of the offence deserves, and the consequentialist rationale, which aims to change the behaviour of the offender or other potential offenders. In practice these two rationales are often combined: the overall size of the sentence can be determined and limited by desert, while the content can be designed to rehabilitate and to reduce the probability of further offences.[7] It appears that social, contextual and psychological information

[5] Raynor, P. 'Three Narratives of Risk: Corrections, Critique and Context' in Trotter, C. et al. eds. *Beyond the Risk Paradigm in Criminal Justice* (Basingstoke, UK: Palgrave, 2016); and Raynor, P., 'Is There Any Sense in Social Inquiry Reports?' (1980) *Probation Journal* 27(3) 78–84.

[6] Pardo, M. and Patterson, D. *Minds, Brains, and Law* (Oxford: Oxford University Press, 2013).

[7] See Asp, P. 'Previous Convictions and Proportionate Punishment under Swedish Law' in Roberts, J. and Von Hirsch, A. eds. *Previous Convictions at Sentencing* (Oxford: Hart Publishing, 2010), 207–226.

can be relevant to both types of sentencing and can influence our perceptions of blameworthiness, although there is often room for argument about what sort of influence is appropriate. For example, do previous convictions increase the seriousness of the current offence or simply reduce the mitigation from which a first offender would expect to benefit?[8] A useful discussion of moral and legal responsibility in different circumstances is provided by Kennett,[9] who sets out to construct a 'common sense moral psychology' by examining the differences between behaviour arising from recklessness, weakness and compulsion.

Only compulsion negates responsibility and needs to be demonstrated by convincing evidence. The reckless person is responsible for failing to take reasonable care and for not resisting an impulse that could have been resisted; the weak person may be unable to exercise control in the present, but can be criticised for failing to take actions in the past that could have been taken, and that could have avoided the current situation in which weakness leads to offending. The role of drink and drugs in the sentencing exercise described earlier would fall under this heading. Self-control can be 'synchronic', exercised in the present, or 'diachronic', exercised in the past to avoid problems later. In either case 'our ascriptions of psychological capacities and competences, including the capacity for self-control, do not require or gain justification from the truth or falsity of some general proposition in physics or neurophysiology'.[10] It is clear that many offenders have difficulties with self-control and/or difficult circumstances in which to try to exercise it. Social and psychological information at the point of sentencing can help to throw light on both these conditions, without negating responsibility except in cases of compulsion.

Predictable Offending

One other way in which neuroscience might be held to threaten our everyday concepts of free choice and responsibility concerns the predictability of offending. If observation of brain states allows us to identify reliable precursors of offending, so that we can predict with a fair degree of accuracy that further offending will occur, does this mean that the

[8] Von Hirsch, A. (2010) 'Proportionality and Progressive Loss of Mitigation: Further Reflections' in Roberts, J. and Von Hirsch, A. eds. Ibid at 1–16.

[9] Kennett, J. *Agency and Responsibility* (Oxford: Oxford University Press, 2001).

[10] Ibid., 164.

choice to offend is predetermined and therefore not free, so that no one could be held responsible for it? This is, of course, simply a variant of the argument considered earlier in this chapter, that caused choices cannot be free. However, it also raises some interesting questions in relation to predictability. The second major use of social and psychological information in criminal justice, in addition to its use in sentencing, is in the assessment of offenders' needs and likely behaviour. A substantial literature on the subject, supported by research on hundreds of thousands of subjects, has grown up over the last 30 years. Research and practice have shown that knowledge of a few key facts about an offender can enable a prediction about the probability of future offending that will be accurate at a much better than chance level. Strictly speaking it is a prediction of the proportion of a group of similar offenders that will offend rather than a prediction about an individual, but it still yields a risk score, usually in the form of a percentage, indicating the probability for an individual member of the group.

The earliest successful reconviction predictors in Britain were developed in the early 1990s by statisticians who used regression analysis of criminal records to calculate the probability of future offending by establishing statistical relationships with age, sex, previous offending and the nature of the current offence.[11] These are known to criminologists as 'static' risk factors because they cannot be altered retrospectively at will. They also tell us nothing about what might need to be changed to reduce the chance of further offending. For this we need a different approach known as 'risk/need' assessment, which was developed mainly in Canada and began to be adopted in Britain in the mid-1990s.[12] This approach mixes some static risk factors with assessments of the social circumstances and personality of the offender, which are mostly 'dynamic' risk factors, also known as criminogenic needs, and are in principle subject to change. Many forensic psychologists now look at what are known as the 'Central Eight' risk factors:[13] these are, in brief, (1) a history of offending, (2) procriminal attitudes, (3) procriminal associates, (4) antisocial personality pattern, (5) adverse family or marital circumstances, (6) problems

[11] Copas, J. B. 'Statistical Analysis for a Risk of Reconviction Predictor' (1992) *Report to the Home Office*. University of Warwick: unpublished.

[12] Raynor, P. 'Risk and Need Assessment in British Probation: The Contribution of LSI-R' (2007) *Psychology, Crime and Law* 13(2) 125–138.

[13] Bonta, J. and Andrews, D. *The Psychology of Criminal Conduct* (London: Routledge, 2017).

around school or work, (7) substance abuse, and (8) absence of prosocial leisure or recreational activities. This kind of assessment provides both an estimate of the risk of reoffending and guidance about what would need to change to reduce the risk, which is helpful for those involved in offender management or rehabilitation. Levels of accuracy achieved by these methods, if properly applied, are typically within a few points either side of 75 per cent correct prediction, with greater accuracy available for very high or very low scorers (for example, people with adverse circumstances, including drug abuse, and a long history of frequent offending are very likely to continue, whilst people in middle age leading broadly satisfactory lives with no history of offending are quite unlikely to start).

There are interesting arguments in the literature about the proper use of risk assessment: for example, parole boards are rightly concerned about who might be too dangerous to release, but extended sentences for high-risk offenders that exceed the normal proportionate sentence for the current offence have been frequently criticised. Risk/need assessment can be used to inform more effective rehabilitation services rather than simply to calibrate levels of coercion.[14] However, it has never been suggested by developers or practitioners of these methods that the fact that an offence can be predicted means that it is not subject to choice. What is being predicted is how that choice will be exercised. Advocates of risk assessment methods have not usually regarded them as deterministic: some of their most determined critics[15] have argued that they can produce an expectation in the minds of professionals, which leads to self-fulfilling prophecies, turning some children into 'crash test dummies' being accelerated toward disaster, but even they do not regard this as inevitable, and they suggest a number of strategies to avoid it. The emergence of new prediction methods based on neuroscience, if this should happen, would not alter the fact that what is being predicted is a choice to offend. In Flew's terms this is 'moving' not 'motion'. It should also be noted that as risk/need assessment is designed to support rehabilitation and to help people to reduce their offending, it actually aims to falsify its own predictions. Far from predetermining offences, it seeks to inform intervention to prevent them.

In some circumstances we have other less-technical ways of predicting offences. For example, we might notice that an offender with a history of

[14] Raynor (2016).

[15] See Case, S. and Haines, K. *Understanding Youth Offending* (Cullompton: Willan, 2009).

violence is becoming tense, breathing hard, speaking loudly, going red in the face, scowling and clenching his fists. It is not hard to make a prediction based on these indicators. Presumably observation of brain states, if the subject would permit this, would yield some similar precursors of violence. For Kennett, this would be a case in which self-control is very difficult to exercise because the current impulse to violence is overwhelmingly strong, but the subject might be held responsible for failing to exercise diachronic self-control before allowing himself to become so angry or so aggressive. He would bear some responsibility for the situation unless he could be shown to have a genuine inability to resist impulses. This inability would have to be observable in a range of situations, not just in the current predicament, and would raise questions about incapacity and potential need for treatment. The relevant difference, for Kennett, is between being incapable of self-control and having a capacity for self-control, which on this occasion is not being used. In practice, some people can find themselves in this kind of state without being aware of it, or not aware until it is too late to regain control. The anger management training that is offered to some offenders is designed to help them to recognise the signs of becoming angry and aggressive and to apply various self-control techniques such as relaxation and appropriate self-talk to interrupt the process and regain control. Kennett argues that the exercise of self-control is primarily a cognitive process, and it is interesting that one of the most successful recent developments in the rehabilitation of offenders has been the development of offending behaviour programmes designed usually by psychologists and based on cognitive-behavioural principles. It is worth considering what might be learned from these about behavioural change, coercion and choice.

Programmes, Learning and Autonomy

Offending behaviour programmes[16] are structured and replicable sequences of exercises and other work carried out with offenders (usually in prison or on probation) to help them to understand and control their behaviour. They are usually designed by psychologists, based on social learning theory and using cognitive-behavioural methods, and when implemented correctly (which they often are not), they typically result, on average, in reductions in reoffending of about ten percentage points

[16] Hollin, C. and Palmer, E. eds. *Offending Behaviour Programmes* (Chichester: Wiley, 2006).

below predicted levels, or below levels achieved by similar offenders dealt with in different ways.[17] In Britain they mainly take the form of group programmes, and when delivered in the community, attendance is required by a court order, though in practice attendance has sometimes been poor. The skills that programmes aim to teach are typically problem solving, interpersonal skills, self-management of emotion, appreciation of other people's point of view, and making decisions by stopping and thinking rather than on impulse. Variations have been developed for special groups of offenders such as those with drug or alcohol problems, sex offenders, domestic violence offenders and others, but virtually all have the same rationale in social learning theory and the same approach to systematic design supported by manuals to ensure that they are implemented as faithfully as possible.

Probation staff reactions to the introduction of programmes were mixed. Some welcomed them, while others were put off by the apparent concentration on thinking skills when they preferred to see crime as a consequence of social disadvantage. Some disliked the standardisation inherent in putting large numbers of people through the same programmes, and more recent designs have included more adaptability to individual offenders. Many of the early reservations about programmes were gathered together in book form.[18] A few probation staff disliked the idea of programmes altogether, perhaps seeing them as similar to brainwashing or to the fictitious coercive therapy inflicted on Alex in Stanley Kubrick's film of Anthony Burgess's novel *A Clockwork Orange*[19]: in other words, they saw programmes as a process that limited and reduced offenders' freedom, in effect reducing crime by creating incapacities. Whilst temporary incapacitation is among the proper goals for custodial punishment, it did not fit with traditional concepts of probation, which in those days was an agreement entered into between the probationer and the sentencing court. (The requirement of consent to a probation order was not abolished until 1997.[20]) However, the experience of programme members who made progress during programmes seemed mainly to reflect the designers' social learning goals. For example, two

[17] See Lösel, F. 'The Efficacy of Correctional Treatment: A Review and Synthesis of Meta-Evaluations' in McGuire, J. ed. *What Works: Reducing Reoffending* (Chichester: Wiley, 1995).

[18] Mair, G. ed. *What Matters in Probation* (Cullompton: Willan, 2004).

[19] Burgess, A. *A Clockwork Orange* (London: Heinemann, 1962).

[20] See Raynor P., 'Consent to Probation in England and Wales: How It Was Abolished, and Why It Matters' (2014) *European Journal of Probation* 6(3) 296–307.

of the graduates of an early programme in Wales made the following comments to a researcher:[21]

> It's helped me to solve problems and get them through to people better. Prevents me from jumping off the handle. I listen more, I think about problems more and discuss things. It takes a lot of stress off my head because I listen to others.
>
> It's made me realise ... It's learnt me to put myself in other people's places if they'd been burgled ... Guilty's the word. It's out of order. It's opened my mind and I look at a subject from all different angles ... Not just jumping the gun. With problems I can clear them up more easily'[22]

Similar comments were made by many of the 64 programme members interviewed. Here we see examples of reduced impulsiveness, more systematic problem solving with awareness of alternatives, more appreciation of the feelings of victims and other people generally, and an improved ability to seek and use help from others. Reduction of impulsiveness, which can be understood as strengthening of self-control, is one of the most common findings from evaluations of offending behaviour programmes. The quotations do not seem to come from people who feel their freedom has been reduced: on the contrary, they show an awareness of more capacity to encounter life and choose how to lead it. They also raise another kind of question about responsibility: if people who lack self-direction, self-control and relevant knowledge get into trouble, they should logically be seen as less to blame then people who have acquired these skills. Perhaps (other things being equal, which they often are not) the successful programme graduate is more blameworthy for the next lapse than somebody who has not had the opportunity to learn responsible behaviour. In other words, some of the mitigation that might be applied to the offender who has limited social competence may not be available to somebody who has learned to have more choice about offending. Knowledge can enable choice, and freedom of choice implies accountability for choices.

Desistance and Choice

One final branch of contemporary criminology that has implications for our understanding of choice and responsibility is the emerging field of

[21] Lucas, J., Raynor, P. and Vanstone, M. *Straight Thinking on Probation One Year On* (Bridgend: Mid Glamorgan Probation Service, 1992).
[22] Ibid., 29.

desistance studies.[23] Desistance theorists aim to throw light on the process of giving up offending, based mainly on qualitative studies of offenders' own accounts rather than on the literature of 'what works' and effective intervention. Crime emerges from desistance studies neither as a simple product of the environment nor as a choice made free from environmental pressures, and this accords with our everyday experience that we do make choices, but we often do not choose the circumstances in which we have to make them. Other insights from desistance studies, which have had some effect on the way criminal justice practitioners approach rehabilitation, include evidence[24] that for persistent offenders, desistance is a process not an event; it is multi-causal; it involves obstacles, mistakes, relapses – 'a zig-zag process', and it often requires development of social and individual 'capital'. It is complete when a noncriminal identity and way of life are firmly established ('primary' becomes 'secondary' desistance) and is facilitated by social bonds to prosocial others, including partners, and parental responsibility. It often requires access to help and motivation to make use of it.

The relationship between environmental influences and individual choice is often expressed in social science as a relationship between structure and agency, and agency appears to play a significant role in desistance. Two studies illustrate this particularly clearly. In a study of released prisoners in Canada, the researchers concluded that established patterns of thinking, and in particular beliefs about their own capacity to overcome obstacles, played a significant role in distinguishing those who stayed out of trouble from those who reoffended:

> In the case of criminal behaviour, factors in the social environment seem influential determinants of initial delinquency for a substantial proportion of offenders . . . but habitual offending is better predicted by looking at an individual's acquired ways of reacting to common situations.[25]

[23] See, for example: Farrall, S., Hunter, B., Sharpe, G. and Calverley, A. *Criminal Careers in Transition: The Social Context of Desistance from Crime* (Oxford: Oxford University Press, 2014).

[24] Summarised in: Maguire, M. and Raynor, P. 'How the Resettlement of Prisoners Promotes Desistance from Crime: Or Does It?' (2006) *Criminology and Criminal justice* 6(1) 19–38.

[25] Zamble, E. and Quinsey, V. *The Criminal Recidivism Process* (Cambridge: Cambridge University Press, 1997) 146–7.

Another influential study is based on interviews with a group of probationers.[26] The researcher summed up his findings as follows: the desisting offender had a 'strong sense that he or she is in control of his or her destiny. Whereas active offenders . . . seemed to have little vision of what the future might hold, desisting interviewees had a plan and were optimistic that they could make it work'.[27]

From this kind of study it appears that desisting offenders see themselves as more free and as having more choices than continuing offenders, who are likely to see themselves as victims of circumstances. The more self-directing desisters are not best understood simply as conforming, but as choosing and building a nonoffending identity and status. This is reminiscent of the condition of 'orthonomy'[28] in which people are both autonomous and seek to choose actions that can be justified according to accepted standards of conduct.

Conclusion

I have argued in this chapter that although recent developments in neuroscience may add considerably to our understanding of human actions, they cannot sufficiently support arguments that deny the existence of choice or responsibility. In other words they cannot undermine or deny our direct experience of human actions. If the hard determinist claims were true, our concept of human action would be fatally compromised in a way that would deny the possibility of an action and fundamentally question our experience of what it means to be human. In short, as Antony Flew put it, 'determinist presuppositions cannot disprove' familiar differences between voluntary and involuntary movements, and between choosing to do something and being coerced. In ordinary usage, doing something of my own free will usually means simply that I was not coerced. Similar arguments advanced by Kenny and Strawson pointed out that our whole vocabulary of communication about our own actions and other people's, including blame, resentment, approval and all similar concepts, would become meaningless if determinism ruled out individual responsibility.[29]

[26] Maruna, S. *Making Good* (Washington, DC: American Psychological Association, 2001).
[27] Ibid., 147. [28] Kennett (2001).
[29] A. Kenny *Freewill and Responsibility* (London: Routledge, 1978); and Strawson, P. F. 'Freedom and Resentment' in Strawson, P. F. ed. *Studies in the Philosophy of Thought and Action* (Oxford: Oxford University Press, 1968).

In addition, it is not clear how deterministic claims arising from neuroscience are new or original. Flew's particular targets were deterministic psychiatrists such as Karl Menninger, who believed that offending was not a matter of choice but was psychodynamically determined,[30] a product of 'unconscious mental forces'.[31] The work of behaviourist psychologists such as B. F. Skinner, who had little interest in 'unconscious mental forces', has also been used as the basis for deterministic arguments against free will and responsibility (as the title of Skinner's best-known book implies[32]). Arguments based on observed brain states appear not to differ fundamentally from the earlier rehearsals of similar arguments: the empirical basis for identifying the brain states may be more convincing than that for 'unconscious mental forces', but it is not clear (except in cases of properly diagnosed mental illness or incapacity) that they are any more capable of eliminating choice and responsibility than other methods of predicting the risk of offending that were discussed earlier in this chapter. A better understanding of the antecedents of our actions, or even of our likely future actions, does not mean we have no control over them or accountability for them.

There is therefore little doubt that neuroscience leaves our common-sense and ordinary-language use of responsibility and culpability largely intact. However, some circumstances that commonly occur in criminal justice call for a realistic view of responsibility based on how far people could reasonably have been expected to exercise control over their behaviour. Offenders who are popularly seen as fully responsible for their harmful behaviour ('bad') often see themselves as victims of circumstances with little choice ('unlucky'). A realistic view usually lies somewhere in between. For example, some offenders may have lost control of their actions in the present, but can reasonably be held to account for failure to exercise control in the past, which would have interrupted the sequence of events that led to the current offence, if opportunities for such control (Kennett's 'diachronic' self-control) did in fact exist in the past. In addition, much rehabilitative work with offenders is directed toward increasing their control over their own behaviour: for example, reducing impulsiveness by teaching people to think before acting (an example of Kennett's 'synchronic' self-control).

[30] Menninger, K. *The Crime of Punishment* (New York: Viking Press, 1968).

[31] Blum, G. S. *Psychodynamics: The Science of Unconscious Mental Forces* (Belmont, CA: Wadsworth Publishing, 1968).

[32] Skinner, B. F. *Beyond Freedom and Dignity* (New York: Knopf, 1973).

Self-control can also be understood as self-direction or autonomy: rehabilitation often increases an offender's freedom of choice rather than aiming to reduce it.

Finally, the challenge of neuroscience may not be catastrophic for our commonsense moral thinking, but it certainly suggests that criminologists should think more systematically about questions of responsibility. In the many cases where offending is made more likely by restricted opportunities not to offend, or by a lifetime subject to antisocial influences beyond the offender's control, some criminologists will see social causation as removing responsibility from the individual offender, as if responsibility must rest either with the individual or with society. However, the arguments reviewed in this chapter suggest that responsibility is often shared, as Kennett argues in relation to disadvantaged offenders. The history of disadvantage does not simply mitigate, it suggests that society's expectation that the offender should avoid further offending needs practical support, both in relation to the individual offender through opportunities and resources for desistance and in relation to potential offenders in general through social policy. Rehabilitation can be society's contribution to support the desisting offender, in a specific instance of the more general social contract between the state and its citizens.[33] Most criminologists' advocacy of rehabilitative and reintegrative justice can only gain from a realistic view of responsibility for offending. In holding offenders to account, we should also be aware of the responsibility of government and other social institutions when their choice of policies increases or fails to address avoidable criminogenic risks that make people more likely to choose to offend. As Kennett puts it, 'holding someone responsible, to whatever extent, need not and should not rule out compassion for their plight or an acknowledgement of any contribution made by society to their predicament.'[34] Current policy often fails this test.

[33] Raynor, P. and Robinson, G. Rehabilitation, *Crime and Justice* 2nd ed. (Basingstoke: Palgrave Macmillan, 2005, 2009).

[34] Kennett (2001) at 198.

10

Neuroscience and the Criminal Jurisdiction: A New Approach to Reliability and Admissibility in the Courts of England and Wales

JOANNA GLYNN

Introduction

This chapter examines recent measures introduced in England and Wales to address issues of reliability concerning complex scientific evidence adduced in criminal trials. For those involved in the development of neuroscientific techniques with potential forensic applicability, these measures will determine whether evidence based on their work will be admissible in criminal proceedings. The Law Commission of England and Wales undertook a study of expert reliability problems in criminal cases and ways of addressing them, including scrutiny of the 'Daubert' test applied in the US. The Commission published its report and draft Criminal Evidence (Experts) Bill in March 2011, proposing a major shift away from the laissez faire 'reliability goes to weight, not admissibility' approach, to a new enhanced reliability test, applied by the trial judge. Subsequently the government rejected almost all the Commission's recommendations. This chapter considers the extent to which the October 2014 amendments to Part 33 (now Part 19) of the Criminal Procedure Rules and the linked Practice Direction have nevertheless given effect to the draft provisions. Despite some uncertainties, it is clear that the degree of scrutiny of scientific methodology and reliability at admissibility stage will be significantly higher than it has been hitherto. Those wishing to introduce 'cutting-edge' neuroscientific evidence in criminal proceedings will now be on notice of the type of information needed to demonstrate reliability; this can only raise standards and facilitate the safe use of important new scientific evidence.

The Need for a New Approach to Reliability and Admissibility in England and Wales: The Problems Identified

On 14 October 2014 the Lord Chief Justice of England and Wales delivered a speech in which he alluded to the increasing pace of change and complexity of scientific techniques underpinning some expert evidence in criminal jury trials. He said that a failure to reform the rules of admissibility, so as to ensure that only evidence based on sufficiently reliable science is placed before juries, will undermine public confidence in forensic science 'with highly undesirable consequences, resulting in either less use of forensic evidence, or less use of juries.'[1]

Currently neuroscience feeds only rarely into the criminal justice system in England and Wales. Whilst we follow with interest developments in neuroscience and their interface with the law, in particular brain-based lie-detection techniques as they develop in the US,[2] evidence founded on these techniques has not been admitted in this jurisdiction. However, as these and other brain-based techniques evolve and become more reliable, this chapter takes the opportunity to explore the new rules of admissibility of expert evidence, in force from 6 October 2014, and how the courts are now likely to approach the admissibility of expert opinion based on complex science. In order to explain these reforms, some reference must be made to what went before, and to lessons learned from the United States.

The criminal courts have most commonly encountered neuroscientific evidence in cases in which specialists in neuroradiology, neuropathology and sometimes neurosurgery provide their opinion as to the cause of an injury, based on their findings on examination of the brain, in vivo or post mortem. These cases usually involve baby deaths or serious injury, described by the acronyms 'NAHI' (nonaccidental head injury') or 'SBS'

[1] Lord Chief Justice, 'The Kalisher Lecture', 14 October 2014.

[2] For example, in *United States v Semrau*, No 7 –10074, M1/P, 2010, the court ruled, applying the *Daubert* test (see para. 5.2 below), that a particular form of brain-based lie detection was insufficiently reliable to be admitted in evidence. However, some commentators think that this test will become forensically useful. (See John Danaher, 'The Future of Brain-Based Lie Detection and the Admissibility of Scientific Evidence' *Irish Criminal Law Journal*, Vol 21, 4, 2011 p67 at p70; and John Danaher 'The Comparative Advantages of Brain-Based Lie Detection: The P300 Concealed Information Test and Pre-Trial Bargaining,' *International Journal of Evidence and Proof*. See also The Royal Society's *Neuroscience and the Law*, 2011 p17, which shows a chart of the type of cases in which neurological or behavioral genetics evidence has been introduced in the USA in criminal cases on behalf of a defendant between 2004 and 2009.)

(shaken baby syndrome'). A number of other specialist areas of medicine, in addition to those that focus on the brain and spinal cord, will provide evidence, each one contributing synergistically to the evidential basis on which a jury may be invited to convict or acquit a defendant of the appropriate offence. However, the primary focus of controversy regarding reliability in this context has been, and remains, the field of neuropathology. Those with an interest in the scientific issues will find them described succinctly in the Furness Report of 2009.[3]

The dangers of expert evidence adduced in emotive criminal cases are well recognised. In 2004 the authors of the Kennedy Report into SUDI (Sudden Unexpected Death in Infancy) cases observed,

> Unfortunately, doctors are occasionally drawn into error because they base their testimony on medical belief rather than scientific evidence. There is also the temptation, particularly in the very adversarial arena of the criminal courts, to be pushed into certainties where there are none . . . just as lawyers and judges can experience case hardening, so can doctors. Those regularly involved with child abuse can find it hard to be dispassionate and indeed sometimes become hawkish. A doctor can be convinced, based on his or her experience, that a defendant is guilty. . .[4]

More generally, there is a volume of evidence to the effect that expert evidence is trusted by tribunals of fact like no other category of evidence.[5] Although not restricted to opinion evidence, an expert, uniquely, may provide his or her opinion, including sometimes on the central issue in the case. The tribunal of fact, usually knowing little or nothing about the

[3] 'Report of a Meeting on the Pathology of Traumatic Head Injury in Children' ['The Furness Report'] December 2009. 'This meeting was convened by the Royal College of Pathologists . . ., in recognition that there had been disagreement between Fellows of the College on the appropriate interpretation of various aspects of the postmortem findings in the cases of traumatic head injury in children.' It is of note that when the Attorney General undertook a review of 297 cases over a ten-year period following the quashing of the conviction in *R v Cannings* [2004] 2 Cr App R 63 (a SUDI case), 97 were cases of 'shaken baby syndrome' (see *Henderson* [2010] EWCA Crim 1269, para. 1).

[4] '*Sudden unexpected death in infancy: A multi-agency protocol for care and investigation: The report of a working group convened by The Royal College of Pathologists and The Royal College of Paediatrics and Child Health.*' Chair: Baroness Helena Kennedy QC.

[5] See Law Commission's Report ('Expert Evidence in Criminal Proceedings in England and Wales,' 2011 Law Com No 325) presented to Parliament on 21 March 2011, alluding to responses to their consultation paper to this effect from the Criminal Bar Association and the GMC.

field on which the expert gives evidence, is reliant on it. This applies as much to the judiciary as to a jury.[6]

As if this is not enough, there are additional concerns about a jury's response to complex scientific expert evidence, so that by 2011 the Law Commission, whose report is considered below, stated (in the context of scientific evidence) that: '... there is a danger that juries will abdicate their duty to ascertain and weigh the facts and simply accept the experts' own opinion evidence, particularly if the evidence is complex and difficult for a non-specialist to understand and evaluate.'[7]

The cases in which convictions have been quashed due to failings in expert evidence are infamous and the details too voluminous to rehearse here. They include, perhaps most notoriously, *Dallagher* [2002] EWCA Crim 1903 (ear-print evidence), *Clark* [2003] EWCA Crim 1020, (SUDI/statistics), and *Cannings* [2004] EWCA Crim 1 (SUDI). In infant death and injury cases, a flavour of some of the problems of expert reliability is to be found in the judgement of Lord Justice Moses in the joined appeals of *Keran Henderson and others* [2010] EWCA Crim 1269 and in the judgment of Mrs Justice (now Lady Justice) King in the Family Division in (*A Local Authority v S* [2009] EWHC 2115). (Both cases arose in the context of alleged baby-shaking cases in which neuropathology was under the spotlight.)

Those judgements describe a small number of neuropathologists, whose evidence was found to have been disingenuous, unreliable, ill-informed, and lacking scientific rigour as a consequence of zealotry and scientific prejudice. One had 'permitted her convictions to lead her analysis.' Whether caused by unwitting confirmation bias or otherwise, the evidence described by Mrs Justice King was clearly unreliable. Those examples relate to expert evidence given to a Family Division judge or to 'fresh evidence' given in the Court of Appeal, Criminal Division, rather than to a jury. However, due to the absence of an effective judicial gatekeeping role until October 2014, unreliable evidence of this sort has been adduced in jury trials, and the jury has been expected to follow the points on reliability, often based on very complex science, and to make decisions of crucial importance without any assistance, beyond that of the judge (who might have come anew to the science at the same time as the jury).

[6] '*... the dependence of the Court on the skill, knowledge and above all, the professional and intellectual integrity of the expert cannot ... be over-emphasised*'. [Wall J. in *Re AB* (Child Abuse: Expert Witnesses) [1995] 1 F.L.R. 181 at 191C].

[7] The Law Commission (2011), para. 1.9.

Although the regulation of expert witnesses and training of all involved in this type of case must play their part, the principal problem was identified by the Law Commission in 2011 as the laissez-faire common law rules of admissibility, which allowed almost any expert evidence to be adduced before a jury as long as it was not *obviously* unreliable.[8]

The Common Law Rules of Admissibility of Expert Evidence

The rules of admissibility of expert evidence in England and Wales have been, and still are, primarily founded on the common law, with some codification in Part 33 (now Part 19) of the Criminal Procedure Rules and several statutory provisions of comparatively little consequence.[9] The common law may be described under four headings, the fourth being the most important for these purposes:

(a) Assistance: Expert opinion evidence is admissible where it provides the court with information that is likely to be outside the experience and knowledge of a judge or jury. [*Turner* [1975] QB 834 and 841]
(b) Relevant expertise: The party seeking to call the expert evidence must demonstrate that the expert is an expert in the relevant field. It must be shown that the individual 'has acquired by study or experience sufficient knowledge of the subject to render his [or her] opinion of value' [*Bonython* [1984] 38 SASR 45 at 47].[10]
(c) Impartiality: The expert must be able to provide impartial, objective evidence on the matters within his or her field of expertise [*Toth v Jarman* [2006] EWCA Civ 1028, [2006] 4 All ER 1276]. This common law admissibility requirement has been reinforced in r.33.2 (now r.19.2) of the Criminal Procedure Rules 2014 (and earlier versions), which provides that an expert has an overriding duty to give opinion evidence that is objective and unbiased.[11]

8 Ibid., 1.8, 1.17, 2.16.

9 For example, s.127 Criminal Justice Act 2003.

10 Guidelines for determining expertise have been developed for certain scientific fields: *Henderson and others* [2010] EWCA Crim 1269, [2010] 2 Cr App R 24 at [207] and [208] (medical experts and the importance of being in practice and up to date) and *Weller* [2010] EWCA Crim 1085 at [36–39] (experts on DNA – jury evaluating the reliability of an expert's opinion entitled to take into account the expert's experience and unpublished work).

11 The fact that a witness was a member of the team that investigated the offence being tried (being on secondment to the Serious Fraud Office for that purpose) did not disqualify him as an expert witness, although it might go to the weight to be attached to his evidence: *R v Gokal* [1999] EWCA Crim 669, CA, Judgement delivered 30 March 1999

(d) Evidentiary reliability (the most important for these purposes)

The case law stresses two points: (i) there is no '*enhanced* reliability test', only a 'reliability test' for the admissibility of experts' opinion evidence; this 'ordinary' reliability test remains undefined, nor does the case law provide guidance on how the courts should assess it;[12] (ii) and 'issues of reliability go to weight rather than admissibility, thus removing any significant gatekeeping role from the judge'.[13] (It should be noted that the term *reliability* here means that the evidence can be relied upon to make a decision, rather than in the scientific sense of repeatable across a number of experiments.)

Illustration in *R. v I, R and T*

The significance of these two points in practice is illustrated by a decision of the Court of Appeal, after the publication of the Law Commission's draft Bill. In *R.v I, R and T* [2012] EWCA Crim 1288; Crim. L.R [2012] 886,[14] the prosecuting local authority appealed against a ruling by the trial judge that expert evidence was inadmissible in the prosecution of the respondent company and its directors for failing to process blood by heating it to the required temperature (133 degrees centigrade) before disposing of it as fertiliser. The fertiliser was tested using counterimmuno electrophoresis (CIE), which tested for the presence of animal proteins. The prosecution relied on the presence of avian protein in the fertiliser, which would not have been found if the blood had been heated to 75 degrees centigrade. The difficulty arose that bacteria are killed if the blood is heated to 125 degrees centigrade, and no bacteria were found. It was common ground that CIE testing had never before been applied to

(hearing 11 March 1999). Similarly, an employee of one of the parties is not necessarily barred from giving expert evidence: *Dunne v Kier Group* [2000] C.L.Y. 326 and *Field v Leeds City Council* [2001] C.P.L.R. 129.

12 *R. v Reed and Reed*, [2009] EWCA Crim 2698 – (case concerning DNA evidence) confirmed a common law reliability test, at least for 'expert evidence of a scientific nature'; but the court did not demur from the established position that *there is no enhanced reliability test* for such evidence. The court did not explain how the trial judge is to determine whether the scientific basis of an expert's evidence is or is not sufficiently reliable to be admissible, save for the hope that the establishment of the Forensic Science Advisory Council and the Forensic Science Regulator will assist the court in its assessment (para. 111). The Court referred only in passing to the Law Commission Consultation Paper of 2009 and the *Daubert* 'enhanced test' for admissibility (see para. 5.2 below).

13 Law Commission (2011) supra n.5 1.11.

14 Judgement of Stanley Burnton LJ.

blood that has purportedly undergone processing in, storage at and onward distribution from a plant such as the defendants'. No similar plants had been tested to compare results following a CIE test. The defendants submitted that although this was a reliable test, it was being applied in a novel context without an evaluation of its efficacy in that context. The trial judge identified the question that forms the basis for the test of admissibility, and then applied it:

> Is the underlying science sufficiently reliable to be admitted in a court of law? The simple answer to that question in this case is 'yes', but in my judgment the better answer is yes unless there are factors over and above the vague and fanciful that cast doubt on the reliability of the CIE test in the context of/on the special facts of this case.

He held that there were such doubt-casting factors.[15]

The Court of Appeal agreed that the prosecution's expert had failed to explain the absence of bacteria in samples showing the presence of animal proteins. However, in allowing the prosecution's appeal, the Court found that the trial judge had confused the question of admissibility with whether, if admitted, the expert evidence was sufficient to gain a conviction:

> In our judgment, once the judge concluded, as he did, that 'the CIE test is a well-recognised and reliable test for establishing whether or not animal

[15] Including: . . . test results which are on the face of it inherently inconsistent with the CIE test, but consistent with the defence case. Some samples tested positive for animal protein, but negative for bacteria. Unless some factor other than heat above 120 degrees centigrade is causing the bacteria to be absent a doubt must remain as to the reliability of the CIE test in what is for it a novel context. The burden must be on the P to prove, by adducing reliable scientific evidence, the 'other factor' in order to confirm the reliability of the CIE test. Mr Reaney postulates the following as being reasons why no bacteria are present in samples which contain animal protein:

1. Dilution of the sample by processed blood thereby reducing the bacteria in the sample to below the detectable threshold.
2. There were no detectable bacteria in the unprocessed blood in the first place.
3. A chemical agent such as disinfectant has killed the bacteria in the unprocessed blood.

None of these hypotheses have been tested. Any one may be right. But Mr Reaney conceded that 2 is probably fanciful. My suspicion is that the truth is No. 1, but my suspicion is neither here nor there. The evidence is that they are all capable of being tested. If they were tested and one or more proved to be the explanation for the inconsistency then there would be no doubt in my mind as to the reliability of the CIE test in this case. Whilst these theories remain untested there is in my judgment a small but reasonable doubt about the reliability of the CIE test results' [judgement paras. 15 and 16].

> protein is or is not present in blood', the CIE test results were admissible on the basis of Mr Reaney's expert evidence. The fact that the test was used for a new purpose, or in a new context, did not of itself render the test unproven ... when the judge proceeded further, to lay down as the test to be applied: are 'there ... factors over and above the vague and fanciful that cast doubt on the reliability of the CIE test in the context of/on the special facts of this case?' The judge confused the question of admissibility with the question whether, if admitted, the expert evidence was such that on its basis a jury could properly convict I.
>
> (judgment paragraphs 28–30).

It can be seen that the threshold for admissibility, applying the 'ordinary reliability test' was low, and the courts regarded what might be described as the 'enhanced' issues of reliability as going to the weight of the evidence, to be assessed by the jury or other fact-finder, rather than going to admissibility. There seems little doubt that the trial judge's approach had been influenced by the Law Commission's draft Bill (considered later) and perhaps the Court of Appeal's decision would have been different, too, had the appeal taken place after the ensuing 2014 changes to the Criminal Procedure Rules (also considered later).

The Criminal Procedure Rules – 'A Novel Way of Implementing an Excellent Report'[16]

A system that can require a jury to make decisions based on unreliable scientific evidence is in need of reform. This part of the chapter explores the measures taken recently to remedy the position and seeks to provide an assessment of the current law on the admissibility of developing areas of scientific evidence.

The Law Commission's Report

The Law Commission undertook a substantial research project, resulting in its Report, 'Expert Evidence in Criminal Proceedings in England and Wales,' 2011 Law Com No 325 presented to Parliament on 21 March 2011, along with its Draft Criminal Evidence (Experts) Bill. As we have seen, traditionally the common law has provided most of the rules of

[16] Lord Chief Justice, 'The Kalisher Lecture', 14 October 2014, para. 17.

admissibility of expert evidence.[17] However, the Commission took the view, based on a significant volume of evidence, that the reliability requirement in the common law admissibility test needed to be reformed by way of primary legislation and not just by guidance from the Court of Appeal.[18]

The Commission scrutinised *Daubert v Merrell Dow Pharmaceuticals* 509 US 579 [US Supreme Court, 1993] as the root from which a detailed test of reliability of *scientific* evidence developed in the US (the 'Daubert test', codified in rule 702, Federal Rules of Evidence).[19] The test assigns to the judge a gatekeeping role, his duty being to ensure that an expert's testimony rests on a reliable foundation and that it is relevant.

The test, usually referred to as 'the enhanced reliability test', is applied in all federal courts and seeks to address the knotty and ambitious problem of providing judges with the tools to distinguish, in terms of admissibility, between science and pseudoscience. The core elements of the test are that the evidence is founded on a testable theory or technique; has been subject to peer review; has a known error rate with standards in place to control its use; and is based on generally accepted science. American commentators have suggested that *Daubert* has

17 Para. 3.1 above. Major changes affecting root and branch aspects of admissibility of evidence are reserved for legislation. For example, Part 11 of the Criminal Justice Act 2003, which brought in major changes to the rules admissibility of hearsay and bad character evidence.

18 The Law Commission (supra n.5 2011) cited judges and practitioners with a wealth of experience, all of whom expressed a view that there was a need for radical reform of the rules on admissibility For example, at 1.23, Bruce Houlder QC, Director of Service Prosecutions: 'My own practitioner's experience . . . is that some charlatans or certainly biased and even incompetent experts still exist in the field of science and also in forensic accountancy. The decisions of the Court of Appeal that have underlined the independent role of the expert, and where their prime duty lies, have not always been heeded, and a 'market' still exists for opinions that assist the cause of the paymaster, and insufficiently scrutinise the value of the evidence that points away from the conclusions contended for at trial.' See also at 1.25, Lord Justice Leveson: 'It is, in my opinion, perfectly clear that expert evidence of doubtful reliability may be admitted too freely with insufficient explanation of the basis for reaching specific conclusions, be challenged too weakly by the opposing advocate and be accepted too readily by the judge or jury at the end of the trial. In that regard, therefore, the law of England and Wales is not satisfactory and reform is undoubtedly required.' See also, Lord Justice Moses in *Henderson and others* [2010] EWCA Crim 1269, [2010] 2 Cr App R 24 at [206]: 'We shall say no more about admissibility since the unsatisfactory state of the law has been the subject of the Law Commission Consultation Paper No 190 . . . and is likely to lead to changes in the current approach of laissez-faire . . .'

19 See the Commission's Consultation Paper No. 190, paras.4.41–4.49.

spawned a huge and expensive new subject of litigation, left many procedural and substantive questions unanswered, and given cause for concern about lack of judicial scientific knowledge and training, inappropriate judicial encroachment on highly contested factual issues, and the assessment of the credibility of expert witnesses and their scientific theories.[20]

Nevertheless the Law Commission proposed the introduction of an enhanced reliability test, explaining that its version had taken into account the lessons learned from the US and demonstrating in its Report the differences between it and the *Daubert* test, both in content and in the method of implementation. Those headline differences are (1) the introduction of guidance and practical training for the judiciary; (2) the policing of the application of the reliability test by the Court of Appeal, who would review a trial judge's decision, not on the basis that he had a broad discretion, which could only be overturned if "*Wednesbury* unreasonable",[21] but on the basis that his ruling was a matter of legal judgment, to be assessed anew according to whether or not it was correct; and (3) by giving the trial judge the power to appoint an expert to assist him in assessing reliability in cases involving exceptionally complex science.[22] The Commission recognised that regulation alone would not

[20] Law Commission Consultation Paper No 190 Part 4, in particular 4.50; and see Margaret Berger, Trustee Professor of Law, Brooklyn Law School, contributing author of the Federal Judicial Centre's 'Reference Manual on Scientific Evidence' Third Edition 2011, p. 36.

[21] The '*Wednesbury* test' is used when assessing the exercise of a judicial discretion as to the admissibility or exclusion of evidence in a criminal trial (from *Associated Provincial Picture Houses Ltd* v *Wednesbury Corporation* [1948] 1 KB 223, 229). It means that, so long as all relevant factors were considered and all irrelevant factors disregarded, the trial judge's ruling will be upheld unless it was a decision no reasonable judge would have reached. Sometimes when assessing a challenge to judicial discretion the Court of Appeal will simply ask itself whether or not the judge's ruling was 'plainly wrong'.

[22] See the Law Commission's Report (2011, supra n.5): 5.91–5.94, 5.110–5.115. 'Our recommendations are positively different in a number of ways from the approach used in jurisdictions that apply the 'Daubert' test' [5.110]. These differences include that: Judges will have the necessary guidance to determine evidentiary reliability; that the Judicial Studies Board will provide trial and appeal judges with practical training in how to assess the reliability of expert opinion evidence in practice; 'the Court of Appeal will properly police the application of the test, by approaching its assessment of the trial judge's performance not on the basis that it is the exercise of a broad discretion governed by 'Wednesbury principles', but on the basis that the judge's ruling is a judgement to be assessed according to whether or not it is right. This will encourage a more critical approach to expert opinion evidence at first instance, with reference to the relevant criteria in the draft Bill, and it will ensure proper scrutiny on appeal'. In addition the

address the reliability issues; the Association of Forensic Science Providers and the Forensic Science Regulator supported the Commission's proposals, seeing them as working synergistically with the quality control measures being taken by them.

The draft Criminal Evidence (Experts) Bill, published with the Law Commission's Report in March 2011 codified the changes advocated by the Commission. It provided, at clauses 1 and 4, for a new enhanced reliability-based admissibility test for expert evidence, whether tendered by the prosecution or the defence, and whether in the Crown Court or the Magistrates' Court. Under the enhanced test expert opinion evidence would not be admitted unless it was adjudged to be *sufficiently reliable* to go before a jury. The test would not need to be applied routinely or unnecessarily. Clause 4(2) lists the reasons for determining that expert opinion evidence is not sufficiently reliable:

(a) the opinion is based on a hypothesis that has not been subjected to sufficient scrutiny (including, where appropriate, experimental or other testing), or which has failed to stand up to scrutiny;
(b) the opinion is based on an unjustifiable assumption;
(c) the opinion is based on flawed data;
(d) the opinion relies on an examination, technique, method or process that was not properly carried out or applied, or was not appropriate for use in the particular case; and
(e) the opinion relies on an inference or conclusion that has not been properly reached.

In assessing reliability the court would have to have regard to such of the generic factors in Part 1 of the Schedule to the Bill as seem relevant and anything else that appears to the court to be relevant. The generic factors cover such considerations as the quality of data on which the opinion is founded, the safety of inferences on which the opinion relies, the accuracy of testing or measuring, peer review of materials, the extent to which the opinion relies on material outside the expert's own field, completeness of information, the range of expert opinion and whether his methods follow established practice.

Clause 5 provided that on appeal the Court was to make its own determination of whether the evidence was sufficiently reliable to be

Commission is recommending 'a facility which would allow the trial judge to call upon additional expertise to assist him or her in the determination of evidentiary reliability in exceptionally complex cases'.

admitted (rather than reviewing the ruling under the Wednesbury principles),[23] while Clause 9 provided for experts appointed by the court to help the court to determine whether the expert evidence is sufficiently reliable to be admitted.

The Government's Refusal to Legislate

The government's response to the Law Commission's report was published on 21 November 2013; most of the Law Commission's 22 recommendations were rejected:

> The Government shares the Law Commission's concern about problems caused by the use of inappropriate or unreliable expert evidence, and is persuaded of the benefits of taking action on this. It recognises the potential value of the proposed reliability test in reducing the risk of unsafe convictions arising from unreliable expert evidence. However, there is no robust estimate of the size of the problem to be tackled – either in terms of the number of cases where unreliable expert evidence is adduced, nor in the impact this has in terms of subsequently quashed convictions. The impact assessment published with the Law Commission's recommendations indicates that application of the new test would involve additional pre-trial hearings, with the concomitant additional costs, but without sufficient reliably predictable savings to compensate for those costs. Without certainty as to the offsetting savings which might be achieved, when set against current resource constraints it is not feasible to implement the proposals in full at this time.

It is clear that the government's refusal to legislate was founded on fears about additional costs and what may be summarised as the absence of a robust risk-benefit analysis.

The decision was met with disappointment by many informed commentators.[24] Cost had been considered by the Commission in the context of almost every proposal, and in particular in Part 6, which deals with court-appointed experts to assist judges assess reliability in cases

[23] For the *Wednesbury* principles see footnote 21. See The Law Commission Report (2011) at para 5.19. 'We believe that the assessment of evidentiary reliability in respect of matters which are not case-specific, principally questions of underlying scientific methodology, should be addressed anew in the Court of Appeal (if leave to appeal is given) not according to whether the trial judge acted within the parameters of a wide discretion' [5.91].

[24] For example Dr Ann Priston, OBE, 'In short, then, the Law Commission's well considered and researched proposals, supported by scientists and lawyers alike . . . have . . . been rejected, primarily on the grounds of cost. What price justice?': Chartered Society of Forensic Sciences' website, accessed in November 2014.

involving exceptionally complex science (draft bill clause 9), where there were undoubtedly cost implications. The Commission emphasized that recourse to the appointment of such experts would be exceptional; clause 9 received support from a wide and impressive cohort of those who responded to the consultation paper, one of whom thought that such a measure, even if rarely used, would act as a deterrent against 'casual science' and might reduce costs in the long run 'as the market for dubious science falls.'[25]

Amendments to the Criminal Procedure Rules

In fact, the government's refusal to legislate has resulted in a tightening of the common law test for reliability, but by an unusual route. The government's response included a reference to inviting the Criminal Procedure Rule Committee to consider amendments to the Criminal Procedure Rules, Part 33.

Previously the courts have recognised the importance of the Criminal Procedure Rules as a tool to ensure that information relevant to an assessment of expert reliability was before the court. In the judgement of the joined appeals on so-called shaken baby syndrome,[26] the Court of Appeal rehearsed the principal provisions of the Criminal Procedure Rules, using the overriding objective of the Rules (at 1.1; namely, to deal

[25] Law Commission's Report (para. 5 above) paras 6.2–6.3, 6.26–6.26, 6.28. The Law Commission's recommendations did have their detractors. Some thought that the stricter application of the common law principles could have done the job. Tony Ward, 'Expert Evidence and the Law Commission: Implementation without Legislation?' (2013) *Criminal Law Review* 561 at 570 noted, 'The advantages of the Law Commission's proposals lie in their clarity, in the occasion that new legislation would afford for judicial training (though the cost of this exercise may well put the government off implementing them), and perhaps most importantly, in the provision for independent experts to advise judges on admissibility decisions.

... The Law Commission report does not envisage a draconian clampdown on the many forensic techniques whose reliability is less than certain; rather it aims to prevent juries from being deceived about how certain the results of those techniques really are. But that is a result that the courts should have been able to achieve – and still could achieve, if the report remains on the shelf – by a strict interpretation to the existing law'. It has also been suggested by one author that the existing law, strictly applied, would be preferable to implementation of the Law Commission recommendations: Adam Wilson 'The Law Commission's Recommendation on Expert Evidence: Sufficient Reliability?' (2012) 3 *Web JCLI*.

[26] *R v Keran Henderson and others* [2010] EWCA Crim 1269, paras. 209–211.

with criminal cases justly) to support the following hitherto not widely recognized procedural 'requirement':

> In the context of Part 33 we should draw attention to the fact that defence experts are not obliged to reveal a previous report they have made in the case, still less to *reveal adverse criticism made by judges in the past.* But a failure to do so will not avail the defence. A judge may well be able to exercise his powers under the Criminal Procedure Rules to ensure that in advance of a trial a defence expert has made disclosure of any relevant previous reports and any adverse judicial criticism. Failure to do so would be contrary to the overriding objective and will achieve no more than to expose the expert to cross-examination on those points at trial. It is difficult to see how those acting on behalf of the defendant could permit an expert report to be advanced without satisfying themselves that previous reports have been disclosed and any adverse judicial criticism identified and disclosed.[27]

Government's response to the Commission's report allowed a more direct approach. And, so, amendments to Part 33 of the Criminal Procedure Rules became the primary vehicle for implementing the recommendations of the Law Commission. The principal features of the October 2014 changes are

- An expansion of experts' duty to the court to include drawing the court's attention 'to any question to which the answer would be outside the expert's area or areas of expertise' (r.33.2(3), now r.19.2(3))
- The introduction to r.33.3 (now r.19.3) of a requirement on the party serving an expert report to 'serve with the report notice of anything of which the party serving it is aware which might reasonably be thought capable of detracting substantially from the credibility of that expert.' (r.33.3(3)(c), now r.19.3(3)(c))
- A new requirement for the content of the expert's report to 'include such information as the court may need to decide whether the expert's opinion is sufficiently reliable to be admissible as evidence' (r.33.4(h), now r.19.4(h))

[27] This suggestion, targeting the defence because the prosecution is already under an obligation to disclose such material, is predicated on the assumption that the defence team is able to access information about previous adverse judicial criticism. In the author's experience, this information is not readily available, even using appropriate internet research tools, particularly if the expert comes from a foreign jurisdiction and s/he has yet to establish a 'reputation'. It was further noted that: 'A case management hearing may often present an opportunity for concerns as to previous criticism of an expert and an expert's previous tendency to travel beyond their expertise to be aired. Whilst such history may not be a ground for refusing the admission of the evidence, it may well trigger second thoughts as to the advisability of calling the witness'.

It is interesting to note that the Practice Direction 33A (now 19A) of October 2014 establishes an unequivocal link between the Practice Direction and the draft bill. It sets out the relevant factors for a judge to use when evaluating the quality and contents of an expert's report in almost identical terms to those of clause 4 and Part 1 of the Schedule to the draft Bill, save that consideration of the Generic Factors in the Schedule to the draft Bill was mandatory; it is not under the Practice Direction.[28]

The Lord Chief Justice, as head of the Rule Committee and responsible for the Rules and the Practice Direction, summarised the changes introduced by the rules and the Practice Direction as follows:

> ... although the common-law remains the source of the criteria by reference to which the court must assess the admissibility, the Rules simply list those matters which must be covered in the experts' report so that the court can conduct such an assessment. The Practice Direction lists the factors the court may take into account in determining the reliability of expert opinion ... This is plainly a novel way of implementing an excellent report.'[29]

Areas of Uncertainty

What is obviously missing from the new Practice Direction are two of the provisions to which the Law Commission attached some importance if the problems encountered in jurisdictions applying the *Daubert* test were to be avoided. The first is the provision allowing the judge to appoint an independent expert to assist him or her in assessing reliability in cases involving scientific issues of exceptional complexity (clause 9 of the draft Bill). There has always been a common law power for a judge to call a witness of fact; although there is no case law on whether that extends to expert witnesses, the Law Commission received evidence that it has been done, albeit rarely. Anticipating that this provision might not be brought into force, they saw no reason why the judge should not call an independent expert pretrial when determining reliability even without

[28] See also Leveson P, *R.v Hamilton* [2014] EWCA Crim 1555 (22 July 2014), para. 44: 'When these changes occur, a new and more rigorous approach on the part of advocates and the courts to the handling of expert evidence must be adopted. That should avoid misunderstandings about what is (and what is not) appropriately included in an expert's report and so either avoid, or at least render far more straightforward, submissions on admissibility such as those made in this case'.

[29] 'The Kalisher Lecture', 14 October 2014, paras. 15 and 17.

the provision; clause 9 simply provided clarity, transparency, a mechanism for appointment, and avoided any inhibition felt by the judge in an adversarial system of appearing to enter the arena; it probably would also have shortened the process, compared to the exercise of the common law power.[30] However, although providing the judge with a statutory power might not have represented a radical change in the law, it remains to be seen whether judges will take such a step in the absence of statutory authority and a known mechanism to do so.

The second missing provision is that relating to reconsideration on appeal, which would have provided the Court of Appeal with the ability to police the operation of the reliability test by determining whether the trial judge's ruling was correct, not whether it was *Wednesbury* unreasonable (clause 5 of the draft Bill). It is not clear how the Court of Appeal will approach appeals under the current law.[31]

The Current Position – *R. v I, R, and T* – Revisited

When the trial judge in *I. R and T* made his decision to rule the prosecution's expert evidence inadmissible, the Law Commission's draft bill had been published for some months; he had clearly applied a test that would require the expert to explain the reliability of his methodology when applied in a novel setting, which was more rigorous than was required in English law at that time. By revisiting that case an assessment of the changes can be made.

It seems clear that the Court of Appeal today would endorse a trial judge's decision in a case similar to that of *R v I, R and T* to undertake prehearing consideration of the factors in the Practice Direction at 33A.5 (now 19A.5) (i.e., the Generic Factors set out in the Schedule to the draft Bill) and 33A.6 (now 19A.6) (i.e., the matters in clause 4(2) of the draft Bill). But how would the Court of Appeal treat the ruling itself?

As we have seen, it is not clear whether the Court of Appeal, applying the current law, would approach an appeal against a judge's ruling on expert admissibility by looking at it anew and determining whether it is wrong in law, or by assessing whether the ruling was '*Wednesbury* unreasonable'. As the facts of this case demonstrate, the distinction could be important; it seems unlikely that the trial judge's ruling would be

[30] Law Commission Report, paras. 6.7–6.8, 6.13, 6.51–6.52, 6.72.

[31] The Law Commission considered whether there should be provision for an interlocutory appeal against a reliability ruling, but decided against it: Report para. 5.98.

deemed '*Wednesbury*' unreasonable, but the Court of Appeal might take a different view to that of the trial judge when considering the underlying scientific methodology of the CIE test afresh, even taking into account the fact that the test had never previously been used for the purpose for which it was deployed by the prosecution, and the expert's failure to explain the findings that did not sit well with the prosecution's case.

A detailed examination of the evidence in *R v I R and T* would not be helpful for present purposes. Whatever the outcome, a student of the changes can be confident that the approach of the Court of Appeal, applying the factors translated from clause 4(2) and the Schedule Part 1 of the draft Bill, would be significantly different after October 2014, and that the trial judge's approach would not be dismissed as the product of a confusion between weight and admissibility.

New Neuroscientific Techniques – Conclusions

The generic challenges in applying neuroscience to the law were explored in the Royal Society's 'Neuroscience and the Law' in 2011.[32] That study concluded that 'discoveries in neuroscience will not completely revolutionize the theory and practice of the law in the near future, but there are already some important practical implications of recent neuroscientific discoveries, which should impact on the law, and there will be many more over the next few years'.[33]

By their very nature, these developments will involve complex science, and reliability will be in issue. If the justice system is to benefit from these new techniques, the science must be usable in court. The introduction of the new reliability-based admissibility factors puts experts, laboratory technicians, persons producing data and others on whose work the expert relies, along with those funding the development of the techniques,[34] on notice of the type of information that will be needed to enable the trial judge, and indeed the parties, to assess the

[32] Brain Waves, module 4, chapter 3, page 10. [33] Ibid., Summary at p. iii.

[34] Issues of funding are outside the scope of this chapter. However, with the unrelenting pressure on public funding of any sort and the closing of the Government's Forensic Science Service about 3 years ago, the cost of developing the techniques in question will inevitably fall to the private sector. It is plain that if the scientific basis for a technique is to be demonstrated to be reliable, all the scientific and technological work, databases and other information that goes to the reliability of the technique will have to be disclosed and made available to all, regardless of issues of commercial confidentiality; see the Lord Chief Justice, 'The Kalisher Lecture', 14 October 2014, para. 28.

reliability of the expert's opinion evidence. This means that detailed information can be prepared toward the end of the development stage, to ensure that the expert can demonstrate that the science is robust enough to withstand challenges to its reliability based on the elements in PD 33A.6 (now 19A.6) (derived from draft clause 4(2) of the Law Commission's Bill). The generic factors in PD 33A.5 (now 19A.5) (derived from the Schedule to the Bill), can be used as a checklist for experts and lawyers alike. Other measures, such as the production of guidance or educational materials about the relevant techniques, published by or in consultation with an appropriate independent body, might also be considered.

The failure to implement two of the Law Commission's recommendations, resulting in uncertainty as to whether the trial judge can appoint an expert to assist him in making the decision as to reliability in cases involving very complex science, and as to how the Court of Appeal will approach its review of the judge's ruling, is regrettable. However, financial constraints affect every part of the criminal justice system, now more than ever, and there are grounds for welcoming the new Rules and Practice Direction as they stand as an improvement on their predecessors. As the Lord Chief Justice concluded his lecture on 14 October 2014,

> through a combination of improving the quality of the scientific underpinning of forensics, the quality, probity and focus of experts, and the manner in which expert evidence is presented to and tested by the courts, we can work with these challenges, and ensure proper safeguards for defendants, while avoiding the risk of juries and the wider public losing faith in the value of forensic science to the courts.

11

Should Individuals with Psychopathy Be Compensated for Their Fearlessness? (Or How Neuroscience Matters for Equality)

MARION GODMAN

Introduction

The science of psychopathy and related conditions continues to put pressure on policy and our legal system. Individuals diagnosed with psychopathy according to the Hare Psychopathy Checklist are disproportionately violent compared to other psychiatric groups. At the same time, their capacity to judge the moral character of their actions is open to question. Although many philosophical arguments about psychopaths' moral psychology merely stipulate certain features conditionally, independent of whether these character traits are ever instantiated[1] – this is not a luxury which the law and our policy makers can afford. Insofar as our legal and political institutions are concerned with the psychological capacities of individuals who actually stand before them, they must also, in some way or another, respond to scientific findings about the nature of those very capacities.

So far the discussion about the psychological capacities of psychopaths has almost exclusively been directed to issues of responsibility and accountability. Should we excuse individuals with psychopathy given that relevant work in psychology and neuroscience indicates systematic deficits in their moral cognition? I will briefly address the issue of responsibility here, but the aim of this chapter is to focus on a somewhat different normative issue: equality. Quite apart from the *responsibility* for the harmful outcomes of violence, how should we

[1] See, e.g., Talbert, M. 'Blame and Responsiveness to Moral Reasons: Are Psychopaths Blameworthy?' (2008) *Pacific Philosophical Quarterly*, 89(4), 516–35; Watson, G. 'Psychopathic Agency and Prudential Deficits' (2013) *Proceedings of the Aristotelian Society*, 113(3), 269–92; Nelkin, D. K. 'Psychopaths, Incorrigible Racists, and the Faces of Responsibility' (2015) *Ethics*, 125(2), 357–90.

distribute the costs of such violence? Should it fall on the psychopathic perpetrator, or do we carry some form of collective burden that may involve, but does not uniquely fall on, the individual actor with psychopathy? I will chiefly focus on this concern for equality with respect to handicaps and tastes as it has been rather famously addressed in the work of Ronald Dworkin, but which so far seems to be missing in the context of psychopathy.[2] This chapter is thus an initial attempt to remedy this gap in the scholarly and public discourse and to encourage further work in that direction.

This issue of equality will be raised in light of some important recent scientific evidence that concerns the excessive use of instrumental aggression in the case of psychopathy. I then go on to argue that despite evidence of a distinctive neural pathway involved in instrumental aggression that detrimentally affects many psychopaths' decision making, this should not be regarded as a grounds for excuse or even diminished responsibility. However, I will argue that it does raise issues about whether the neurocognitive deviation, or as I will sometimes refer to it more colloquially, fearlessness, should be seen as a handicap or as a resource. If it is regarded as a handicap, we enter into a domain of political philosophy, as the costs for this violence should not exclusively fall on the psychopathy – doing so would be unfair to them. But if it is a matter of a *resource* that actually assists the individual in pursuing goals according to their own values, then it is hardly something that justifies compensation.

Two Kinds of Violent Minds

Psychopathy is typically regarded as a subclass of antisocial personality disorder (ASPD). The two disorders overlap in the diagnostic characteristics like impulsivity, failure to plan ahead, recklessness, irritability and, indeed, aggressiveness, as indicated by the frequent participation in physical fights and assaults.[3] Nevertheless an increasing amount of evidence suggests that the groups are doubly dissociable. For example,

2 See Dworkin, R. 'What Is Equality? Part 1: Equality of Welfare' (1981a) *Philosophy & Public Affairs*, 10(3), 185–246 and Dworkin R. 'What Is Equality? Part 2: Equality of Resources' (1981b) *Philosophy & Public Affairs*, 10(4), 283–345.

3 Hare, R. D., and Neumann, C. N. 'The PCL-R Assessment of Psychopathy: Development, Structural Properties, and New Directions' in C. Patrick (ed.), *Handbook of Psychopathy* (New York: Guilford, 2006).

people who have ASPD without psychopathy often have social anxiety, and it is not uncommon for ASPD to be comorbid with mood and anxiety disorders. Such comorbidity is typically absent in the case of psychopathy. Those with psychopathy also have a different interpersonal style than those diagnosed as ASPD, which includes shallow affect, manipulativeness, grandiosity, lack of guilt and remorse, and as a group, they are also more dangerous according to the criminal statistics: individuals with psychopathy begin offending at an earlier age, they engage in a broader range and higher frequency of offending behaviour, and they also respond less well to common current behavioural interventions in childhood and adulthood.

The past decade of work in psychology and neuroscience suggests that these differential patterns of personality and behaviour might be underwritten by two different kinds of aggression: reactive aggression on the one hand, and proactive or instrumental aggression, on the other. In the case of reactive aggression, the aggression is usually understood as a response to a perceived provocation (threat, frustration, etc.). To be sure, the population that scores high on the Hare Psychopathy Checklist also display increased reactive aggression compared to the controls, but it is their excessive use of proactive aggression that sets them apart from people with only ASPD. Proactive aggression in contrast to reactive aggression is controlled and motivated by the expectation of external rewards.[4] An influential study of the role of aggression in children revealed different information-processing patterns characteristic of the two forms of aggression.[5] Whereas a hostile attributional bias was characteristic of reactive-aggressive children, a relatively positive evaluation of aggression was more characteristic of children prone to proactive-aggressive. Although the same individual may of course experience both forms of aggression, this suggests two quite distinct phenomenologies of aggression, with the proactive case appearing not only more voluntary, but also more positively evaluated.

Although there may not be a perfect alignment between crimes that are a result of proactive aggression and those that are a result of reactive

[4] Dodge, K. A., and Coie, J. D. 'Social-Information-Processing Factors in Reactive and Proactive Aggression in Children's Peer Groups' (1987) *Journal of Personality and Social Psychology*, 53(6), 1146–58.

[5] See Crick, N. R., and Dodge, K. A. 'Social Information-Processing Mechanisms in Reactive and Proactive Aggression' (1996) *Child Development*, 67(3), 993–1002; and Vitiello, B., and Stoff, D. M. 'Subtypes of Aggression and their Relevance to Child Psychiatry' (1997) *Journal of the American Academy of Child & Adolescent Psychiatry*, 36(3), 307–15.

aggression, there has been some successful operationalisation in investigations as to whether psychopathy is indeed overrepresented in the predatory and calculated crime statistics. Two different studies found that instrumental offenders did score higher than both reactive and nonviolent offenders on the Hare Psychopathy Checklist.[6] Within a group of 125 Canadian convicts of murder, it was also found that among the psychopaths, homicides that were of an instrumental nature were highly overrepresented (93.3 per cent) in contrast to the nonpsychopathic offenders, where homicides that were instrumental in nature represented about half of the total (48.4 per cent).

A common suggestion for what is characteristic of psychopathy that might also explain the prevalence of instrumental aggression in this group is their comparative *lack of sensitivity to fear conditioning* as compared to reward conditioning.[7] It has also been suggested that the failure to use punishment information to signal inappropriate behaviour is one of the first characteristics of psychopathy, emerging early in childhood.[8] However, the mechanism for this fearlessness in decision making and how it relates to the difference between psychopathy and ASPD has not yet been settled. Do punishments or sanctions that deter normal subjects simply not work in the case of psychopathy?[9] A recent neuroimaging study of reinforcement learning in psychopaths published in *The Lancet* suggests it is not as simple as characterising male

6 See, Cornell, D. G., Warren, J., Hawk, G., et al. 'Psychopathy in Instrumental and Reactive Violent Offenders' (1996) *Journal of Consulting and Clinical Psychology*, 64(4), 783–90.

7 See Blair, R. J. R., Colledge, E., Murray, L., and Mitchell, D. G. V. 'A Selective Impairment in the Processing of Sad and Fearful Expressions in Children with Psychopathic Tendencies' (2001) *Journal of Abnormal Child Psychology*, 29(6), 491–8; Blair, R. J. R., Mitchell, D. G. V., Leonard, A., et al. 'Passive Avoidance Learning in Individuals with Psychopathy: Modulation by Reward but Not by Punishment' (2004) *Personality and Individual Differences*, 37(6), 1179–92; and Birbaumer, N., Veit, R., Lotze, M., et al., 'Deficient Fear Conditioning in Psychopathy: A Functional Magnetic Resonance Imaging Study' (2005) *Archives of General Psychiatry*, 62(7), 799–805.

8 See Blair, R. J. R., Peschardt, K. S., Budhani, S., Mitchell, D. G. V., and Pine, D. S. 'The Development of Psychopathy' (2006) *Journal of Child Psychology and Psychiatry*, 47(3–4), 262–76.

9 One issue here that I think deserves further empirical investigation is if there is a distinctive kind of punishment that is more or less effective. The way punishment is operationalised in the experiments is typically only by some negative signal (red light, etc.). In particular it is not, for example, clear whether social sanctions like ostracisation, shaming etc. that are normally effective in correcting behaviour works the same in the case of psychopathy.

psychopathy as 'diminished neural sensitivity to punishment'.[10] Rather, the results from a probabilistic response-reversal learning tasks show that it is not the sensitivity to punishment per se that is lacking – it is rather that past punishments informs the decision making in atypical ways in psychopathy. The MRI scans of the subjects with psychopathy showed an *increased* response to punished reversal errors in the circuitry of anterior cingulate cortex and anterior insol (areas responsible for tracking the salience of rewards and punishments). The authors take this to indicate a prediction error; roughly speaking, the subjects diagnosed with psychopathy had not anticipated the punishment and were surprised. The result also contrasted with subjects with ASPD, but without psychopathy, and controls without either condition. Yet the anticipation of reward seems to be in concordance with other subjects. This suggests that it is the altered organisation of the information-processing system responsible for reinforcement learning and decision-making patterns that is distinctive of psychopathy. It is not that individuals with psychopathy aren't sensitive to punishment; past punishments or deterrents simply are not as well encoded or are not salient to them in their current decision making.

Now, much of the empirical evidence described in this section awaits both elaboration and replication, but at least we have a hypothesis for a plausible neurocognitive mechanism or differential neural organisation that can explain why instrumental violence is more prevalent in the population of people with psychopathy. The threat of punishment that frequently works as a deterrent for risky behaviour like violence and assault simply has less of a grip on people with psychopathy. The neural variation in the reward and punishment-guided decision making between subjects with psychopathy and subjects with ASPD suggested by these studies describes a way in which neuroscience at least prima facie becomes relevant to the law. It partly confirms the psychologically dissociable proactive and reactive aggression, but it also substantiates and details the mechanism guiding voluntary person-level behaviour more precisely.

Still, what also becomes apparent in the case of violent crime caused by violent aggression is that the problem is not restricted to the aggression itself. Past punishment in relevantly similar situations is regarded as less of an obstacle in pursuit of the ends, but these ends are pursued

[10] Gregory, S., Blair, R. J., Simmons, A., et al. 'Punishment and Psychopathy: A Case-Control Functional MRI Investigation of Reinforcement Learning in Violent Antisocial Personality Disordered Men' (2015) *The Lancet Psychiatry*, 2(2), 153–60.

independently of – but perhaps also in addition to – the predicted violence. Nothing in the neuroscience (or in the psychological studies more generally) so far tells us why – or indeed whether – people with psychopathy pursue the distinctive type of (immoral or violent) ends from us others.

The Responsible Instrumentally Violent Mind

To settle the relevance of the preceding findings for the moral accountability of individuals diagnosed with psychopathy, we have to make a few assumptions. First, we have to suppose that the evidence holds up and can be replicated. In particular it would be nice to identify a distinct neural organisation or mechanism underlying the impaired decision-making distinctive of psychopathy.[11] For example, if it is *not* neural sensitivity per se, but the lack of impact of past punishment, is it due to regions that indicate cognitive deficits or motivational biases? As far as the law is concerned we also need to suppose that we are able to identify and successfully operationalise certain premeditated crimes so that we can establish or isolate the use of instrumental as the proximate cause of violence (rather than, say, reactive aggression). Alternatively, or, *in addition*, we should suppose that the scans of altered brain organisation of individuals with instrumental aggression is reliable enough to be used as probative evidence.

Supposing for a moment that all this is in place and the scientific corroboration and practical obstacles can be overcome, what are the implications? Let us first ask about the implications for responsibility: should we in some circumstances excuse individuals with psychopathy of their violent crimes (for example, by extending the current insanity defense or analogous legal codes)? Or perhaps do individuals with psychopathy at times fall short of the requirement for mens rea, and so does this yield a basis for diminished responsibility?[12] Pardo and Patterson, in their recent book, display a rare optimism toward the prospective

[11] After all, so far the evidence only shows that that such decision-making deficits connected to instrumental aggression are more prevalent in the population of people with psychopathy compared to other clinical groups – not that it is exclusive to that group and that individuals without psychopathy couldn't share the same deficit.

[12] Often the inclusion of psychopathy in the insanity defense is resisted on practical and quasi-ethical grounds: individuals with psychopathy would not want to use the defense anyway. Some reason that it would be unfair and unsafe to people with other psychiatric disorder to place them with psychopaths; there is no way to 'cure' people of psychopathy, etc. In another

use of neuroscience in this domain. Although they claim that this kind of evidence does not challenge any of the existing criminal doctrinal categories (*actus reus,* mens rea, or affirmative defenses, e.g., proof of insanity) per se, they think it might bear on the scope of application of such doctrines. That is, it might be that individuals with psychopathy as a group fall short of *culpable intent* or even fulfill the condition of an insanity defense.[13] In fact, providing evidence in individual cases is in their opinion 'the most plausible avenues by which neuroscience can contribute to the law'.[14]

Although I agree with their general line of thinking about what neuroscience can and cannot contribute, the case of instrumental aggression and the associated fearlessness is not one that falls squarely in providing evidence for excuse or diminished responsibility. The first thing to say is that the use of aggression is merely the most proximate cause of the person's behaviour whether they have psychopathy or not. The whole idea of aggression being instrumental is that it presupposes that there is already an intention to achieve whatever end that happens to be urgent or salient. So if proactive aggression is indeed instrumental in nature what motivates it? Of course, people with psychopathy are not alone in using anger instrumentally. Robert Frank in his seminal *Passions within Reason*[15], argues that anger might in fact be rationally motivated – just like many other seemingly irrational emotions. Such an instance might be avenging a wrongdoing with the expectation that the long-term reputational gains would outweigh any immediate costs to the individual.[16] Frank's hypothesis certainly seems plausible in light of our literary and dramatic canon with vivid depictions of gang violence, male dueling and even war in pursuit of recognition. Arguably, for this very reason,

paper myself and my coauthor argue that none of these concerns are themselves persuasive and so an obstacle for including psychopaths under such legal mechanisms (See, Godman, M., and Jefferson, A. 'On Blaming and Punishing Psychopaths' (2017) *Criminal Law and Philosophy,* 11(1):127–42; see also Nadelhoffer, T. and Sinnott-Armstrong, W. 'Is Psychopathy a Mental Disease?' in Nicole Vincent (ed.), *Neuroscience and Legal Responsibility* (Oxford: Oxford University Press, (2013), pp. 229–55.) Here I want to insist that such grounds for exculpation or diminished responsibility are unjustified even if they were feasible to implement.

13 Pardo, M. and Patterson, D. *Minds, Brains, and Law* (Oxford: Oxford University Press, 2013), Chapter 5.

14 Ibid., 140.

15 Frank R. *Passions within Reason: The Strategic Role of the Emotions* (New York: W. W. Norton, 1988).

16 Ibid. at 83.

foreseeable reputational gains is encouraged by extremist groups and other gangs and cultures of violence. Is reputational gains also behind the use of this violence in psychopathy?[17] Some indications might be gleaned from the studies of psychopathy in violent crimes: for example, psychopaths' victims are much more often strangers, and so it is for example highly atypical for their violent crimes to arise from domestic disputes or from drug abuse.[18] Can this be understood as a suggestion that ceteris paribus individuals with psychopathy are susceptible to the reputational gains? Or could it be that the arousal attached to the emotion of anger might itself be motivating in many instances?

In fact, I am inclined to think that speculating about general or even more prevalent motives within this clinical population of psychopathy might be wrongheaded when it comes to matters of responsibility.[19] Whatever motivates the particular agent, even in the case of psychopathy, it is in some sense their *own* motives and so something we should be reluctant to excuse. We can contrast this with cases where the agent acquires motives that are foreign to them due to a tumor or a delusion. Take what was referred as the '*Samurai case*' (samuraifallet) in Sweden, where a man attacked his girlfriend with his samurai sword suddenly in the middle of the night under the drunken delusion that she was attacking him.[20] Now although we might dispute the court's decision to lessen the sentence (the violence may seem disproportional to the threat and arguably also nonmitigating due to the self-induced intoxication), the intention of attacking his girlfriend seems at least to be caused by a motive foreign to the man. He was deluded. The same counterfactual does not seem to apply to the case of the harm due to instrumental aggression; that is, even if say John, who suffers from psychopathy, was sufficiently moved by past punishments or the threat of future punishments, he would probably still have intended to do what he

[17] See Williamson, S., Hare, R. D., and Wong, S. 'Violence: Criminal Psychopaths and their Victims (1987) *Canadian Journal of Behavioral Science*, 9(4), 455–62. Whether or not reputational gain or prestige is the typical or generic motivation for instrumental aggression *in particular* is contestable. Certainly we should not assume that violence pursued for instrumental ends is morally bad, all things considered. It may, for example, be pursued as strategic self-defense, protest and ultimately lead to more social justice.

[18] This, in turn, is in line with the idea that the violent aggression does not require a perceived provocation, and that the use of alcohol or drugs is not required to engage in highly risky behaviour.

[19] Thanks to Teemu Toppinen for making this point vivid to me.

[20] 'Samurajmålet', Umeå tingsrätt (Umeå district court), NJA 2012 s. 45.

intended to do. Or at the very least it is not obvious that he would not have *wanted* to do what he did. In the samurai case it is highly unlikely that the boyfriend would have attempted to kill his girlfriend if he had not perceived that he was under attack.

As a rejoinder, one might insist that the individual with psychopathy violent aggression is not only explained by prior motives and intentions, but at least partly explained by some *deficits* like the inability to draw on past experienced punishments as deterrents for intended risky and violent behaviour. To be sure, these constraints (or opportunities, depending on how you view them – more on this in the next section), figure in the actual determination of what we decide to do. If I was not fearless of heights, I might not for example be *as* inclined on mountain climbing as I, in fact, am. But although the awareness of one's fearlessness might influence the *strength* of one's motives, it does not seem to determine the motives in the first place. Not even in the case, I would contend, for pure thrill seekers – after all, they don't seek all kinds of thrills all of the time. Similarly, in the case of psychopathy: the fearlessness and lack of impact of past sanctions does not determine that they should want to behave violently, harmfully or immorally in the first place. In fact, if psychopathy is a clinical condition or psychiatric disorder and not simply a designation of bad people or a reflection of our moral disapproval, we should also acknowledge the possibility that much of the time many individuals with psychopathy do *not* want to behave violently toward others.

The Cost and Collective Responsibility for Instrumentally Violent Minds?

In the previous section, I argued that the empirical evidence of the deficits connected to the use of proactive aggression alone do not excuse responsibility. Although individuals with psychopathy may be less deterred by risks and threats of other social sanctions, they still have the freedom to choose which moral gambles to participate in. In the absence of empirical evidence for simply bad motives being associated with the condition of psychopathy, it seems simply patronising to assume that an individual with psychopathy could not want to be or want to do well. Unlike the person suffering from a delusion or hallucinations, whose values and motives are foreign to the agent's overall outlook, there is no evidence so far that the values that inform the individual with psychopathy's intentions and subsequent actions do *not* have the same degree of autonomy as our own. Certainly many of the values may *seem* alien to us, but that is precisely because we regard them from our own, external, perspective.

Yet at this very juncture there seems to me a neglected issue concerning the role of fearlessness and the deviant decision making that is informed by past rewards but not by past punishment. We have concluded that the prediction error in question does not amount to an incapacity that is exculpating or mitigating, but what about its status from a perspective of a society that tries to treat people fairly. The question is this: is the deviant decision making something that we, as a society, share some collective burden or responsibility for? If so, it seems that the agent suffering from this incapacity can simultaneously be held responsible as well as owed compensation by others. This form of question is posed by Ronald Dworkin in his seminal discussion of the meaning of equality.[21] Dworkin considers the issue as related to individual 'handicaps'. Should we redistribute resources to individuals to correct for the outcomes of their handicaps? But then, should we also compensate for their disproportionate needs due to the particular 'expensive' tastes? Dworkin claims our theory should advise us to answer 'yes' to the former, and 'no' to the latter.

According to Dworkin, in an equal society where everyone has an equal share of the resources to pursue what they see as good, handicaps are the type of thing for which the agent deserves some form of compensation. According to Dworkin, this is because they have been acquired due to brute luck: 'they are born with them, develop them before they are entitled to or have sufficient knowledge to insure on their own behalf'.[22] Because they cannot buy insurance after the event of such bad or brute luck, they are entitled to compensation from society for this unfairness. He contrasts this with the case of so-called expensive tastes, which in his view are not the sort of thing that should figure in matters of redistribution or compensation (even though the agent may not have resources enough to pursue their tastes and so as a result suffer from a decline in welfare if they do not receive compensation).[23] This is either because the expensive tastes are preferences that are *voluntarily and knowingly*

[21] Dworkin (1981a and 1981b). [22] Ibid. (1981b), 297.

[23] In a previous article Dworkin argues that equality does not make sense as a target independent of a theory of distributing equal resources. In a nutshell his argument is that any recommended changes in distribution of welfare is self-defeating, as they presuppose some the theory about what resources an individual is entitled to have at her disposal in leading a life at all (1981b).

formed with a sense of the cost to other, or *cravings* that may be acquired by accident but have become intimately connected to who that person is (the example he gives is of someone who has developed generous appetite for sex). In either case, it is unlikely that people would take 'hypothetical insurance' against the tastes or cravings because not having them would imply imagining a different conception of what they want from life than what they, in fact, want. So in the case of voluntary tastes and important cravings, we should not allocate any compensatory resources to enable the agent to pursue their ends.[24]

To me there seems to be some potential tension between these different normative justification Dworkin gives for the distinction between handicaps and expensive tastes: is it the source (luck vs. voluntary acquisition) or the integration with the overall agential outlook?[25] Unfortunately, though, getting embroiled in this debate would take me on too much of a detour for present purpose, and so I will mostly discuss the distinction between expensive tastes and handicaps in terms of the differential *source* involved in its acquisition, with expensive tastes being primarily voluntary, and handicaps involuntary. The key question at hand, then, is to which category does the fearlessness displayed by individuals with psychopathy belong: handicaps or expensive tastes/cravings?

Let's first suppose we can regard the lack of impact of past punishment as a handicap, which individuals with psychopathy have acquired at no fault of their own and which they have been unable to insure themselves against. Following Dworkin, it would then be correct for society to correct or compensate for this unfairness. The fearlessness and lack of impact in past punishment does not excuse, but it entitles individuals with psychopathy to claims of compensation just like other diseases or illnesses do in Dworkin's luck egalitarianism, where there has been no possibility of prior consent, knowledge or means for insurance.

On the other hand, let's suppose we regard the lack of impact of past punishment as a reflection of the individual's 'expensive taste'. If so, (some) individuals with psychopathy would have voluntarily fostered a fearlessness to be able to pursue their preferences for ends that are quite possibly also intimately connected to who they are. Analogously to other strategies adopted in pursuit of one's cravings, it can be argued that the

[24] Ibid. at 320f. Thanks to Simo Kyllönen for helpful discussions of this.

[25] For criticism of luck egalitarianism along such grounds, see Anderson (1999).

individuals in question have purposely cultivated their fearlessness and capacity for instrumental aggression. They identify with the values and simply want to ensure that they have the best possible chances to pursue their particular ends. As I have mentioned earlier, some ends might be external rewards associated with violence, like recognition and prestige, whereas others could be intrinsic rewards associated with violence, abuse and causing harm to others. But whatever these values and preferences that motivate the individual might be, if we think of them in accordance with the model of expensive tastes, the lack of effect that a risk of punishment and threat of social sanctions is an *advantage*, rather than an *obstacle* in their pursuits. In other words, the agent would not have opted for insurance against the particular craving or taste even if there had been an option.[26]

So what kind of evidence might help determine whether we should regard the instrumental aggression of psychopaths as a result of a handicap – and hence amenable to compensation – or as a result of a taste and purposeful fearlessness? One form of evidence that at first may seem to be relevant is how important social influences are to why individuals with psychopaths adopt their particular values and ends in the first place. After all, social values, like reputation, prestige and also social norms for how to achieve them (and for or against the use of aggression), are not equally distributed in all societies. This may have an impact on the distribution and prevalence of psychopathy in different societies. But although there are certainly social influences on *which* values are adopted, the fact that society and culture influences our values per se seems like a fact relatively constant throughout the human social world. Moreover, in the last section I argued that there is no reason to think that the values aren't just as voluntarily (or involuntarily) adopted by people with psychopathy as in nonpsychopathic cases (for similar reasons, it seems premature to claim that psychopathy is a malaise of contemporary society).[27] And equally they seem to be values that the

[26] It might seem like the difference between understanding fearlessness (due to lack of impact of past punishment) as a handicap or an expensive taste roughly parallels the distinction between psychopathy as a mental illness or kind and psychopathy a moral kind; whether they are 'mad' or simply 'bad' (Maibom 2008). However, the distinction here has the advantage of not labeling the individual *as a whole* one way or another but merely the particular trait that the person at any given time may or may not be exhibiting.

[27] Ronson, J. *The Psychopath Test: A Journey through the Madness Industry* (New York: Riverhead Books, 2012).

individual adopt as his or her own. So the social influences on values *neither* exculpates nor warrants compensation.

A more promising place to look for evidence of whether we are dealing with an involuntarily adopted handicap or not is the source of the deviant decision making or fearlessness in the typical developmental profile of psychopathy. Indeed, most people grant that there are plenty of influences on the cognitive traits typical of psychopathy that are beyond the individuals' direct voluntary control.[28] Some might be outside the control because they are (at least partly) genetically inherited. Research, for example, suggests that the long allele of the serotonin transporter confers a risk for psychopathy.[29] In interaction with aversive conditions during childhood (also for the most part outside the agent's control), this might affect the deviant decision making in psychopathy as previously reviewed.[30] Indeed, perhaps research will determine that this particular genetic contribution might be part of the explanation for why these individuals fail to be deterred by past sanctions and punishment, contributing to the distinctive fearlessness of psychopathy. Indulging in some speculation about the acquirement of this involuntary handicap, the developmental story might run as follows: first, due to factors outside the individual's control (probably a combination of genetic factors and aversive childhood experience), social sanctions and other punishments have a reduced role in influencing the decision making of many individuals with psychopathy (at least compared to the normal population and those with only antisocial personality disorder). Direct or indirect punishment (such as blaming or in other ways disapproving) will thus fail to get its typical grip in the decision making. In particular, there is probably less obstacle to pursue ends by means of aggression that would normally have been deterred by past social sanctions. As any particular punishments imposed by the community continuously fail to have much influence on future behaviour, these individuals end up in a vicious circle of bad behaviour with an increased overall likelihood of using anger and violence.

[28] See Blair et al. (2006).

[29] Glenn, A. L. 'The Other Allele: Exploring the Long Allele of the Serotonin Transporter Gene as a Potential Risk Factor for Psychopathy: A Review of the Parallels in Findings' (2011) *Neuroscience & Biobehavioral Reviews*, 35(3), 612–20.

[30] See Reif, A., Rösler, M., Freitag, C. M., et al. 'Nature and Nurture Predispose to Violent Behavior: Serotonergic Genes and Adverse Childhood Environment' (2007) *Neuropsychopharmacology*, 32(11), 2375–83.

Still before we jump to the conclusion that the instrumental aggression and fearlessness in the case of psychopathy is a handicap due to a specific involuntary feedback loop between genes and (social) environment, we should note that the majority of individuals with the longa allele of the serotonin transporter will still not be psychopathic, even if they display some of the endophenotypes associated with the condition.[31] So it could still be that there is also some degree of voluntariness in the development of the course of the condition given that not all individuals with the genetic (and perhaps also, developmental) profile of psychopathy are instrumentally aggressive to the same extent. If so the fearlessness and lack of attention or concern to past punishment/sanctions in correcting future behaviour found in individuals with psychopathy may, in fact, be driven by the demands of whatever values drive actions in the first place. For example, it may still be that to achieve the kind of prestige the particular individual desires, a neglect or lack of concern with the harm and blame associated the instrumental aggression is purposefully adopted. What's more, just like Dworkin suggests in the case of cravings, it may become an instrumental part of the overall outlook of the agent. If so, it looks more like fearlessness and aggression due to an expensive taste than a handicap.

In sum, it seems premature to make any specific judgement as to whether we ought to regard the fearlessness and instrumental aggression present in psychopathy as a handicap or as a requirement due to the agent's expensive taste. What we need is more research into the development of psychopathy (which is often difficult due to ethical reasons) and in particular, we need research on the interaction between genetic and environmental factors in the deviant decision making, fearlessness and use of instrumental aggression. This will naturally also involve neuroscience. What is clear is that there are two putative channels of societal influence in the development of the condition: what values the individual acquires and their potential reinforcement (susceptibility to punishment and rewards). Although I think only the latter might give us grounds for regarding the fearfulness as a handicap, we should note that both channels of influence are something we are beginning to understand are responsive to some societal intervention.

If we were to regard individuals with psychopathy as having a handicap, this is something that we might not only be able to compensate for

[31] Glenn (2011).

materially but also try to prevent by early intervention in the developmental trajectory of the disorder.[32] Although a 'clinical pessimism' about treatment in psychopathy still prevails, work that focuses on changes in social practice like parental interactions with children and adolescents with psychopathic traits has shown some success,[33] as has some that focus on pharmacological interventions to the brain.[34] We should remember that the neuroscience so far indicates that influence via social *rewards* like praise, gratitude, and different means of reciprocation and bonding seem to be able to impact decision making of individuals with psychopathy just as much as in subjects that do not have psychopathy. This suggests that *positive* social relations and feedback, in contrast to negative feedback, may still be effective in encouraging a different set of values than those that typically guide individuals with psychopathy.[35] Whether it is possible to focus solely on this side of the moral education and neglect sanctions and punishment, however, remains to be seen.[36]

One might wonder, of course, whether we *should* intervene on the developing minds in this way, particularly when it comes to what values are adopted.[37] After all a liberal society should certainly not to coercively constrain individuals to adopt the values that the majority happens to endorse or that might lead to narrowing the conception of flourishing. At the same time, our survey into the way the decision making is compromised in the case of psychopathy suggests that these individuals may indeed not have had the equal opportunity when it comes to exploring a variety of different conceptions of flourishing. They have not had an equal chance to learn from punishment, and this

[32] The more probable normative grounds for interventions is our duty to protect against future violence that might be perpetrated by individuals prone to instrumental aggression. Of course the obligation to prevent harm can also be combined with the societal duty to compensate.

[33] McDonald, R., Dodson, M. C., Rosenfield, D., and Jouriles, E. N. 'Effects of a Parenting Intervention on Features of Psychopathy in Children (2011) *Journal of Abnormal Child Psychology*, 39(7), 1013–23.

[34] Glenn (2011).

[35] Kochanska, G. and Aksan, N. 'Children's Conscience and Self-Regulation' (2006) *Journal of Personality* 74(6), 1587–618.; and Godman, M. and Jefferson, A. 'On Blaming and Punishing Psychopaths' (2017) *Criminal Law and Philosophy*, 11(1), 127–42.

[36] Caldwell, M.F. and van Rybroek G. 'Effective Treatment Programs for Violent Adolescents: Programmatic Challenges and Promising Features' (2013) Aggression and Violent Behaviour, 18(5), 571–578

[37] Glannon, W. 'Intervening in the Psychopath's Brain' (2014) *Theoretical Medicine and Bioethics*, 35(1), 43–57.

in fact may have discouraged certain forms of valuable life and biased other forms of life that are harmful to others (and, probably, also to themselves).

Conclusion

Patterson and Pardo argue that current neuroscience should not lead to any major revisions with respect to our central legal doctrines. I agree. I have reviewed a recent set of findings that might go some ways in explaining the prevalence of violent crimes distinctive of psychopathy, namely the use of instrumental aggression due to not being deterred by past sanctions and punishment in current decision making. I argued that this should not lead us to think that individuals with psychopathy can be excused for many of their crimes. So the findings do not challenge our current legal practice and policy with respect to the responsibility of individuals with psychopathy either.

Nevertheless, I have introduced a different challenge posed by such findings. This challenge concerns the possible *collective* responsibility we as a society might have for this compromised decision making of violent individuals with psychopathy due to (1) the possibility that the fearlessness displayed in psychopathy may be regarded as a handicap, i.e., not something the individual has had the knowledge or ability to insure herself against, and (2) our knowledge and ability to intervene on this condition as a form of compensation (or perhaps act of prevention). I'm afraid I have fallen short of taking a determinate stance on whether fearlessness actually should amount to a handicap and be compensated in terms of a social intervention. Instead, my aim has been to begin to shed some light on how the neuroscience and psychology of psychopathy matters to issues of equality and the aim of equal distribution of resources. Subsequently these scientific findings also matter to law and policy making concerned with compensation as well as societal and pharmaceutical interventions.

12

The Treatment of Psychopathy: Conceptual and Ethical Issues

ELIZABETH SHAW

Introduction

If effective treatments were developed for individuals with psychopathic personality disorder, this would produce considerable benefits for society, by saving lives, preventing (potentially irreversible) psychological and physical harm and generating large economic savings. Psychopaths represent less than 1 per cent of the population but are responsible for a disproportionate amount of crime. According to some estimates, up to 15 per cent of the prison population have this disorder.[1] In particular, psychopaths are disproportionately responsible for the most serious offences – including murder and violent and sexual assaults.[2] US researchers have found that the total economic cost associated with psychopaths' criminal behaviour (including the direct costs of the harm they do as well as the sums the criminal justice system spends on prosecuting and imprisoning them) amounts to more than the combined cost to the economy of drug and alcohol abuse.[3] In recent years, in England, authorities have recognised the urgent need to adopt more effective responses to individuals with this condition and have channelled considerable funding into the Dangerous and Severe Personality Disorder Programme (DSPD), which ran between 2000 and 2013 and which has now been replaced by the Offender Personality Disorder Pathway (OPDP). One of the main aims of these programmes is to increase understanding of which interventions are best able to treat and rehabilitate high-risk psychopathic offenders. In contrast, there has been no comparative ring-fenced funding in Scotland, and the development of services for individuals with psychopathy has progressed much more slowly than in England.

[1] K Kiehl and M Hoffman MB. The Criminal Psychopath: History, Neuroscience, Treatment and Economics (2011) 51 *Jurimetrics* 355.

[2] Ibid. [3] Ibid.

There used to be a widespread view among psychiatrists and those working in the criminal justice system that it was impossible to treat psychopathy. In contrast, today, writers increasingly argue that such pessimism is unwarranted. It is true that, so far, although research into treatments for psychopathy has yielded some intriguing findings, there is not yet clear evidence that any particular intervention is effective in treating offenders with this condition. However, the failure to gather reliable evidence about effective treatments may be due to the fact that, until relatively recently, research in this area has been insufficiently funded, and such studies (including research into the effectiveness of the DSPD interventions) faced various methodological problems, which may well be overcome by future researchers. This chapter will not focus on such empirical hurdles. Instead it will consider some key conceptual and ethical questions, which should be considered before embarking on empirical research. First, it will discuss whether there are principled reasons (as opposed to empirical reasons) why it would be futile to attempt to alter psychopaths' problematic character traits or behaviour using biomedical means. Second, it will consider whether there is something about the nature of psychopathy, which means that it could *never* be ethically permissible to offer psychopathic offenders any treatments that potentially carry significant risks. Influential writers on this topic have recently defended both of these contentions. It is essential to consider these objections, because if they were sound, then it would be a serious mistake to devote substantial resources to researching such neurointerventions for psychopaths. Instead, efforts should be redirected into considering the best ways of detaining and managing offenders without such treatments. However, this article will attempt to cast doubt on these objections. It will then propose a number of preconditions that should be satisfied before neurointerventions could permissibly be given to psychopathic offenders.

The Nature of Psychopathy

The standard tool for diagnosing psychopathy is the 'Psychopathy Checklist – Revised' (PCL-R), which is a list of 20 criteria.[4] The presence

[4] The original Psychopathy Checklist was created by R Hare in 'A Research Scale of the Assessment of Psychopathy in Criminal Populations' (1980) 1 *Personality and Individual Differences* 111. It was revised by R Hare, *The Hare Psychopathy Checklist Revised* (2nd ed. Multi-Health Systems, Toronto 2003). Similar checklists for juvenile psychopathy have

of each item on the PCL-R is scored on a scale from 0 to 2 (0 means the trait is absent, 1 means it is possibly or partially present and 2 means it is definitely present). The maximum possible total score for all 20 items is 40. A diagnosis of psychopathy can be given to individuals with scores of 30 or above. This section will only focus on a selection of the most salient characteristics of psychopathy, which might be the target of future treatments. It is important to remember that psychopathy exists on a spectrum ranging from mild to severe. The characteristics discussed in this section may be more or less pronounced depending on the severity of the condition.

Psychopaths are notorious for their inability or impaired ability to feel 'affective empathy'. This can be defined as the capacity to experience an appropriate emotion in response to perceiving another person's emotions (e.g., compassion in response to someone's distress) and an imaginative ability to put oneself in someone else's shoes and feel what it is like to be in their situation. Psychopathy has been characterised by James Blair as a developmental neurological disorder. Psychopathy is associated with structural and functional differences in regions of the brain associated with empathy in nonpsychopaths. For instance, studies suggest that psychopaths typically have reduced neural tissue in the amygdala and orbitofrontal cortex. This lack of affective empathy is associated with a tendency to engage in cold, callous behaviour and with 'instrumental aggression' – the deliberate use of aggression to achieve one's goals. Psychopaths' lack of empathy emerges at a very early age, and the aforementioned brain abnormalities have been reported in children as young as five.[5] The psychologist, Paul Frick, claims that even two-year-old children with impaired empathy may begin to show signs of their impairment through their behaviour. For example, normally developing toddlers will occasionally hit other children. However, they

been developed. These include the P Frick and R Hare, *The Anti-Social Process Screening Device* (Multi-Health Systems, Toronto, Ontario 2001) and A Forth et al., *The Psychopathy Checklist – Youth Version* (Multi-Health Systems, Toronto, Ontario 2003). Features of psychopathy have been divided into interpersonal and affective characteristics on the one hand and lifestyle issues on the other. The former category includes superficial charm, grandiosity, pathological lying, lack of remorse or guilt, manipulativeness, shallow affect, callousness/lack of empathy and failure to accept responsibilities. The latter category includes a need for stimulation, proneness to boredom, parasitic lifestyle, poor behaviour controls, early behavioural problems, lack of realistic long-term goals, impulsivity, irresponsibility and juvenile delinquency.

[5] J Blair et al., *The Psychopath: Emotion and the Brain* (Blackwell, Oxford 2005).

will usually show some signs of distress when they see the harm they have caused (e.g., the other child's tears) and when the parent expresses disapproval. However, children with impaired empathy will often be completely unfazed by other peoples' distress or disapproval or by parental discipline.[6]

Psychopaths' emotions are often described as shallow, and many psychopaths may have a reduced ability to feel fear. In most people, fear and anxiety are accompanied by various autonomic responses, e.g., increased heart rate and sweating of the palms. However, some studies show that psychopaths tend to have reduced autonomic responses to the anticipation of a painful electric shock.[7] Psychopaths' abnormal emotional responses may be one reason why they seem to lack understanding of other people's emotions. Psychopaths often appear puzzled by their victims' reactions. For instance, one psychopathic offender said, 'They are frightened, right? But, you see, I don't really understand it. I've been frightened myself, and it wasn't unpleasant'.[8]

Despite the fact that psychopathy is not associated with low intelligence, psychopaths often engage in strikingly self-defeating behaviour. Psychopaths' offences are often carelessly planned, such offenders are often easily caught and they seem undeterred by past experiences of punishment. For example, Hughes et al. describe one psychopathic repeat offender called Bob, who claimed that his employers' money had been stolen by a mysterious robber.[9] But Bob described the robber's weapon in such great detail that the police became suspicious and found this weapon (which apparently Bob had not tried hard to conceal) at Bob's own residence. A similar example is an 18-year-old psychopath who, rather than admitting his delinquent activities to his father, concocted a story about being injured in a wrestling match and being treated by his family doctor, despite knowing that this lie would be discovered the next day because he was due to visit the same doctor,

[6] Paul Frick, quoted by Jacqueline Stenson, 'Destined as a Psychopath: Experts Seek Clues', www.nbcnews.com/id/30267075/ns/health-mental_health/t/destined-psychopath-experts-seek-clues/#.WPKnao61sXo, last accessed 15/04/17.

[7] R Lopez, R Poy, C Patrick, and J Molto, 'Deficient Fear Conditioning and Self-Reported Psychopathy: The Role of Fearless Dominance' (2013) 50 *Psychophysiology*, 210–18.

[8] R Hare, *Without Conscience: The Disturbing World of the Psychopaths among Us* (The Guilford Press, New York 1999).

[9] M Hughes, M Dolan, and J Stout, 'Decision-Making in Psychopathy' (2015) 23(4) *Psychiatry, Psychology and Law* 521, p. 521.

accompanied by his father for a prearranged appointment.[10] Rather than indicating limited intelligence, Hervey Cleckley argued that that such behaviour is caused by a strange sort of apathy.[11] The psychopath will lie or commit crimes based on a mere whim, but will not take the trouble to avoid being caught.

Psychiatrists' accounts of psychopaths' life histories reveal a chaotic series of reckless actions that, in addition to harming others, typically result in injury and punishment for the psychopath himself. The psychopath's self-defeating behaviour may arise from a general lack of serious, lasting concern about anyone's interests – including his own. According to Hervey Cleckley, psychopaths do not seem to care deeply about anything, but only to act on whims. Cleckley draws an analogy with a man dangling off a ledge, who knows that if he lets go he will be seriously injured. The psychopath is similar to 'someone in this position who, furthermore, is not very tired and who knows help will arrive in a few minutes, but who, nevertheless, with a charming smile and a wisecrack, releases his hold to light a cigarette, to snatch at a butterfly, or just to thumb his nose at a fellow passing in the street below'.[12]

There is evidence that some of the psychopath's strange behaviour could be explained by problems engaging in 'mental time travel' (MTT) – the ability to recall autobiographical events and to imagine possible future scenarios.[13] MTT also has an affective dimension – one identifies emotionally with one's future and past self. This sense of emotional identification may be lacking in the psychopath. Doris McIlwain wrote that when psychopaths imagine their past or future lives, 'it is as if they are rummaging around in someone else's diary and photographs, making plans the way a travel agent plots a trip for someone else.'[14] Problems with MTT could account for psychopaths' characteristic lack of long-term goals and could also explain their impulsive behaviour, as MTT is important for self-control. MTT enables nonpsychopaths to experience in the present moment some of the positive emotions that one will feel on achieving long-term goals. This may reduce the emotional salience of desires to give in to transient temptations that might

[10] H. Cleckley, *The Mask of Sanity* (5th ed. Mosby, St Louis 1988), p. 344. [11] Ibid.
[12] Ibid., p. 392.
[13] N Levy, 'Psychopaths and Blame: The Argument from Content' (2014) 27(3) *Philosophical Psychology* 351.
[14] D McIlwain, 'Living Strangely in Time: Emotions, Masks and Morals in Psychopathically-Inclined People' (2010) 6(1) *European Journal of Analytic Philosophy* 75, 91.

interfere with long-term plans. In contrast, psychopaths seem to be 'stuck in the present', at the mercy of fleeting impulses.[15] Levy argues that problems with MTT may also contribute to psychopaths' apparent lack of empathy for others. He argues that psychopaths' impaired ability to form an integrated sense of themselves, as individuals continuing over time, with goals that help to give meaning to their lives, could make it hard or impossible for psychopaths to imagine that other people have such goals. According to Levy, psychopaths cannot fully 'appreciate that others have *lives*; ongoing plans and projects'.[16]

Having said this, many psychopaths probably have *some* hopes for/concerns about the immediate future (even if they experience these in an emotionally dulled way). For example, if questioned directly about whether they wish to be imprisoned, they would say 'no'. However, their ability to translate even these limited hopes/concerns into action may be impeded by a kind of attention-deficit disorder[17] that makes it harder for them to focus on several different factors simultaneously.

It is unclear whether the characteristics of psychopathy are caused primarily by environmental factors (e.g., childhood neglect and maltreatment) or by genetic factors. It is possible that different individuals may develop the same syndrome by different causal routes. Theorists have debated whether the characteristics of psychopathy mean that individuals with this condition should be relieved from criminal responsibility. However, this chapter will not attempt to settle this question, but instead will focus on the potential for treatments to reduce psychopaths' criminal tendencies.

The Connection between Psychopathy and Crime

Psychopathy is often characterised by the *lack* or *impairment* of certain abilities, such as the ability to empathise, to experience certain emotions, or to care about and focus on one's long-term interests. Psychopathy can therefore lead to the *absence* of certain motives that typically cause people to refrain from crime – motives like compassion and enlightened self-interest. Psychopathy in itself does not necessarily

[15] Ibid. [16] N Levy (2014), p. 363.
[17] J Newman et al., 'Attention Moderates the Fearlessness of Psychopathic Offenders' (2010) 67 *Biological Psychiatry* 66.

involve the positive *presence* of strong motives to offend. However, psychopaths can, of course, experience such antisocial motives (e.g., violent, sadistic urges), even though such motives may not be an essential feature of psychopathy. A psychopath can be compared to car with faulty brakes. If they are heading down a road that will expose them to danger and bring them into conflict with others, then they have an increased risk of causing harm because their faulty 'psychological brakes' may not stop them. However, if they do not have antisocial desires, then it is as if they are heading down a relatively deserted road, and their faulty brakes will not lead to a major collision.

Although psychopaths are disproportionately represented among violent criminals, the majority of psychopaths do not commit violent crimes, and some psychopaths do not commit crimes at all. When psychopaths do break the law their behaviour is often characterised by 'criminal versatility' – a tendency to commit a variety of different offences, rather than specialising in one type of offence.[18] Although good parenting may not be able to prevent the disorder from developing, it may play some role in how the disorder is expressed.[19] If a child with psychopathic tendencies is taught prosocial ways of fulfilling his/her desires and has access to resources to enable her to do so, this seems to reduce the chance that the child will become an offender (at least to some extent). Conversely, poor parenting and social deprivation increase the risk that such a child will commit crimes.

Treatments

It is possible to distinguish between several different aims of potential treatments (a treatment programme might also include a combination of these different aims). First, a treatment might aim to *increase psychopaths' emotional and imaginative abilities*. For example, potential treatments might involve the neurotransmitter, oxytocin, which appears to play a role in empathy. Some studies suggest that oxytocin receptors in the brains of psychopaths are abnormal.[20] However, researchers have expressed caution about the idea of simply

[18] R Hare (1999). [19] Ibid., p. 174.

[20] M Dadds, C Moul, A Cauchi, et al., 'Methylation of the Oxytocin Receptor Gene and Oxytocin Blood Levels in the Development of Psychopathy' (2014) 26(1) *Development and Psychopathology* 33.

administering oxytocin to psychopathic adults or children with psychopathic traits, as the mechanisms involved are complex, and currently they are poorly understood.[21] Second, a treatment might aim to *enhance psychopaths' cognitive abilities.* For example, such treatments might increase psychopaths' ability to focus on the costs of antisocial behaviour when they are tempted to break the law. Research on the use of methylphenidate with offenders suffering from psychopathic and antisocial disorders suggests that this treatment may reduce their impulsivity and inattention.[22] However, this treatment would have to be carefully monitored, as patients can abuse it, and psychopaths are prone to substance abuse.[23] Cognitive enhancements might also increase psychopaths' ability to identify alternative, prosocial means of pursuing their goals. It might be worried that cognitive enhancements could make psychopaths worse – by enabling them to plan violent crimes more effectively. However, this worry might be alleviated by the fact that so-called 'successful psychopaths' seem to be much *better* at exercising certain cognitive skills (such as focus and mental time travel) than 'unsuccessful psychopaths', and yet successful psychopaths are also much *less* likely to be violent.[24] Third, a treatment might aim to *reduce the strength of antisocial urges* that may accompany psychopathy (while not being a symptom of psychopathy itself). For example, there is some evidence that various pharmacological interventions, including selective serotonin reuptake inhibitors (SSRIs), may reduce aggressive impulses.[25]

Treatments can also be distinguished in terms of their level of invasiveness and potential side effects. At the more invasive end of the spectrum is deep brain stimulation (DBS). DBS is a neurosurgical procedure that is currently used to treat neurological disorders such as Parkinson's disease. It involves implanting electrodes in the brain to stimulate certain brain areas. De Ridder et al. have suggested that it may

21 R Mackenzie and J Watts, 'Does Lack of a Moral Feeling Brain Indicate Moral Disability? Children Diagnosed with Callous Unemotional Traits, Emotion Regulation and the Potential of Treatment with Oxytocin' (2012) 17 (4) *Tizard Learning Disability Review* (4) 184.

22 B Dolan and J Coid, *Psychopathic and Antisocial Personality Disorders. Treatment and Research Issues* (Gaskell/Royal College of Psychiatrists 1993).

23 Ibid. 24 Levy (2014) at 351.

25 P Ferari et al., 'Escalated Aggressive Behavior: Dopamine, Serotonin and GABA', (2005) 526 *European Journal of Pharmacology* 51; T Douglas, 'Moral Enhancement' (2008) 25 (3) *Journal of Applied Philosophy* 228; M Crockett et al., 'Serotonin Selectively Influences Moral Judgment and Behavior through Effects on Harm Aversion' (2010) 107(40) *Psychological and Cognitive Sciences* 17433.

become possible to use this technique to correct the structural and functional abnormalities found in psychopaths' brains.[26] The pharmacological treatments mentioned earlier are less invasive and typically significantly less dangerous than neurosurgery. Transcranial direct current stimulation (tDCS) is often described as 'noninvasive'. There is some evidence that tDCS may reduce risk taking.[27] It involves applying electrodes to the scalp and transmitting small electrical currents to the brain. It is not as precise as DBS, as it can only target deeper brain regions by passing through the outer layers of the brain. Although it is generally regarded as a safe treatment, the electrical current does spread from the point of application to brain tissues and adjacent regions (which meets some definitions of 'invasive') and is not completely free of side effects.[28] Neurofeedback is a technique that is clearly noninvasive and does not appear to have side effects. It involves viewing real-time visual images of one's brain activity and retraining one's brain wave patterns in response to this feedback. Farah Focquaert has argued that this is a promising treatment for psychopathy.[29]

It is important to emphasise that biomedical interventions are most likely to be successful if they are *combined* with psychological and behavioural therapies, such as cognitive behavioural therapy (CBT). Umbach et al. have suggested that the possibility of incorporating biological measures into behavioural treatment programmes for psychopaths is an area that deserves more attention.[30] For example, some studies suggest that measures of certain biological factors, e.g., hormone levels, may give some indication of whether an individual is likely to

[26] D De Ridder, B Langguth, M Plazier, and T Menovsky, 'Moral Dysfunction: Theoretical Model and Potential Neurosurgical Treatments' in J Verplaetse, S Vanneste, J Braeckman, and J Schrijver (eds.), *The Moral Brain: Essays on the Evolutionary and Neuroscientific Aspects of Morality* (Springer, New York 2009) 155–84.

[27] S Fecteau, D Knoch, F Fregni, N Sultani, P Boggio, and A Pascual-Leone, 'Diminishing Risk-Taking Behavior by Modulating Activity in the Prefrontal Cortex: A Direct Current Stimulation Study' (2007) 27(46) *The Journal of Neuroscience* 12500.

[28] N Davis and M Koningsbruggen, '"Non-Invasive" Brain Stimulation Is Not Non-Invasive' (2013) 7(76) *Frontiers in Systems Neuroscience* 1–4.

[29] F Focquaert, 'Mandatory Neurotechnological Treatment: Ethical Issues' (2014) 35 *Theoretical Medicine and Bioethics* 59.

[30] R Umbach, C Berryessa, and A Raine, 'Brain Imaging Research on Psychopathy: Implications for Punishment, Prediction, and Treatment in Youth and Adults' (2015) 43(4) *Journal of Criminal Justice* 295–306.

benefit from a particular type of treatment, e.g., cognitive behavioural therapy.[31] In the future, biological measures (including brain-imaging research) might be used to help identify which individuals with psychopathy were most suitable for a particular treatment programme. Furthermore, Umbach et al. suggest that observing biological changes after an intervention like CBT could help to identify the mechanism through which the treatment works.[32]

Treatments of psychopathy are more likely to be successful if they are consensual. Zlodre et al. argue that individuals with personality disorders are more likely to engage in the treatment if steps are taken first to increase their understanding of the nature of the treatment and the need for it and, second, to reduce any feeling of being coerced.[33] However, this is particularly challenging with psychopathic patients. These patients may have a diminished understanding of *what* is being treated, as their condition involves limited insight into their own personality and behaviour. Their condition also undermines their ability to appreciate *why* they should be treated, or to care about these reasons. One reason for treatment is to prevent psychopaths from harming themselves through their risky behaviour. However, psychopaths may not understand or be motivated by this reason, as they arguably have a reduced ability to realise that their behaviour is self-defeating and a diminished ability to care about their own long-term interests. Another reason for treatment is to prevent harm to others. However, psychopathy also involves a lack of concern about the impact of their behaviour on others. Furthermore, psychopaths typically have difficult relationships with clinical staff. This can prevent psychopaths from properly understanding the nature of the treatment (because they are often unwilling to ask staff for clarification), as well as increasing the extent to which psychopaths perceive treatment as coercive.

Despite the fact that psychopathy undeniably gives rise to a number of significant challenges, which can prevent psychopaths from benefiting from traditional treatment approaches, attempts are increasingly being made to tailor treatment programmes specifically to psychopathic patients. For example, the Chromis programme for violent psychopathic offenders

[31] L Cornet, C Kogel, H Nijman, A Raine, and P Laan, 'Neurobiological Changes after Intervention in Individuals with Anti-Social Behaviour: A Literature Review' (2014) 25(1) *Criminal Behaviour and Mental Health* 10.

[32] Ibid.

[33] J Zlodre, J Yiend, T Burns, and S. Fazel, 'Coercion, Competence, and Consent in Offenders with Personality Disorder' (2016) 22(4) *Psychology, Crime & Law* 315.

(whose pilot phase within a prison unit was funded by the DSPD) aims to overcome some of the difficulties outlined earlier. The treatment does not aim to change offenders' personality traits directly, but to give offenders the opportunity to enhance their decision-making skills so that they are able to find prosocial outlets for their desires and avoid/resolve conflict situations. Even though psychopaths may not be strongly motivated to engage in the treatment programme by a desire to achieve long-term goals, they may be more likely to engage if the programme itself absorbs their attention and is rewarding in the short term. Therefore, the Chromis programme 'uses novel and stimulating material in sessions, avoids repetition and employs a wide range of techniques and methods including elements of competition'.[34] Staff on the programme also seek to identify any personal interests and priorities that the patient does have and to take an individualized approach, relating the treatment to these priorities and interests. The programme also aims to decrease the experience of being coerced into treatment. For example, participants are given the opportunity to actively control certain aspects of their treatment, e.g., 'materials can be adapted and even created by individual participants so that they meet their own preferred learning styles'.[35] Although initial evaluations of this programme have reported some promising findings, it is still too early to say whether such programmes will be successful.

The next section will consider the objection that there are principled reasons why treatment programmes that involve biomedical interventions are doomed to failure and will attempt to rebut this objection.

Are There Principled Reasons Why It Is Impossible to Alter Psychopaths' Traits Through Biomedical Means?

Heidi Maibom claims that there are principled reasons why direct brain interventions may never be able to modify psychopaths' psychological characteristics.[36] Maibom suggests that attempts to alter a psychopath's moral and social worldview (or Weltanshauung) by pharmacological means may be doomed to failure because of the unified nature of this Weltanshauung. (Maibom's argument also has implications for

[34] J Tew and R Atkinson, 'The Chromis Programme: From Conception to Evaluation' (2013) 19(5) *Psychology, Crime & Law* 415, p. 420.

[35] Ibid.

[36] H Maibom, 'To Treat a Psychopath' (2014) 35 *Theoretical Medicine and Bioethics* 31.

the general issue of attempting to address the antisocial behaviour of any offender – including offenders who would not be diagnosed as psychopathic.)

As noted earlier, some researchers claim that the psychopath's inability to empathise removes the normal inhibition against harming others. Discussions of developing rehabilitative interventions for psychopaths have often focused on the possibility of enabling them to empathise. The problem with such interventions, according to Maibom, is that the psychopath already possesses an entire worldview that has been shaped by his lack of empathy. Therefore, she claims, pharmacological modification of brain processes connected with empathy will be 'swamped' by the numerous and interlinked psychological traits that make up the individual's preexisting worldview. She writes,

> [A weltanshauung] is a diffuse network of beliefs, assumptions, attitudes, and affective dispositions related to a vision of how to live (including how to live with others).
>
> ... People strive to have reasonably coherent views of the world, and moral considerations must be weighed against each other, and against other considerations. Even if a change were introduced [by the direct brain intervention] in one set of such considerations, it would still be answerable to all the other beliefs, attitudes, etc. that make up the person's moral outlook ... If what is introduced is incompatible with or sufficiently in tension with other parts of the subject's outlook, it will be eliminated.[37]

Maibom's account assumes that the psychopath already has a kind of 'moral outlook', which, presumably, involves beliefs that others do not have moral worth and that it is permissible to harm them. However, it is unclear whether psychopaths do have the host of moral beliefs and attitudes that Maibom appears to ascribe to them. Some writers have argued that those psychopaths with a lifelong incapacity to experience any feeling of caring concern for others are (as a result) unable to grasp moral concepts connected with others' rights or welfare.[38] If this is correct, then psychopathy may actually involve the *absence* of a moral outlook, rather than a *positive worldview* that denies that others have moral worth. This would cast doubt on Maibom's claim that a new sensation of empathy would necessarily be swamped by opposing values that the individual already holds. There is no reason, in principle, why

[37] Ibid., p. 40.
[38] E.g., S Morse, 'Psychopathy and Criminal Responsibility' (2008) 1 *Neuroethics* 205.

the introduction of a new capacity could not be an important factor leading to a radical change in one's worldview. For example, imagine someone who had been blind from birth gaining the capacity to see. This person might develop a passion for visual art, leading her to acquire a wide range of beliefs, attitudes and aesthetic values. More argument would be required to show that instilling the capacity to empathise could not have an analogous effect.

Maibom is correct to highlight the complex interactions between an individual's various attitudes, beliefs and emotions. This complexity makes it unlikely that a direct brain intervention, *on its own*, would instill appropriate values in psychopaths. However, it might provide a basis on which other approaches, e.g., moral dialogue, could get a foothold.

Maibom objects that, because psychopathy is a 'moral disorder', the individual herself must make a moral commitment to engaging in the rehabilitation process before moral change can occur. Maibom claims the psychopath could never make such a commitment, due to the nature of the condition, which is characterized by narrow, short-term self-interest. Yet, it is conceivable that the psychopath might initially enroll in a rehabilitation programme for self-interested reasons (e.g., to increase the chance of early release) and that a moral commitment to continue with it could be built up gradually over the course of the rehabilitation programme, as interventions designed to instill empathy begin to take effect. Furthermore, some studies suggest that even short-term self-interest (e.g., the offer of regular financial incentives for abstaining from drugs) may play a useful role in bringing about long-term behavioural changes.[39] The fact that drug misuse arguably also has a 'moral' dimension does not prevent self-interested incentives from serving this function.

Is It Unethical to Treat Psychopathy, Due to the Nature of This Condition?

This section will consider whether there is something about the nature of psychopathy, which means that it could never be ethically permissible to offer psychopathic offenders treatments for their condition (in particular, treatments that potentially carry significant risks).

[39] J Neale, C Tompkins and J Strang, 'Qualitative Evaluation of a Novel Contingency Management-Related Intervention for Patients Receiving Supervised Injectable Opioid Treatment' (2016) 111(4) *Addiction* 665.

Hubner and White have raised this type of objection.[40] They endorse the general principle that it is unethical to offer a treatment to a patient if that treatment will produce no medical benefit for the patient. They argue that psychopaths do not experience subjective suffering as a result of their condition and that therefore neurointerventions to treat psychopathy will have no medical benefit for these patients. These individuals typically identify with their psychopathic traits and do not regard them as problematic or a source of distress. On the contrary, the 'symptoms' of their disorder may actually give them pleasure. Rather than viewing their disorder as a disability that could be helped by treatment, they may experience a perverse pride in their 'abilities', e.g., their ability to deceive and manipulate, or their ability ruthlessly to pursue their ends, while remaining untroubled by other people's distress. Hubner and White acknowledge that psychopaths may suffer as an *indirect* result of their condition – psychopaths typically do not like the social consequences of their behaviour, such as punishment or preventive hospital detention. Successfully treating psychopaths could confer a kind of benefit on them if it enabled them to be released. However, Hubner and White reject the idea that freedom from external sanctions can count as a 'medical benefit'. Hubner and White argue that if psychopaths were considered fit candidates for treatment, then there would be no reason to refrain from 'treating' anyone who is regarded as a social outsider according to prevailing social norms. They draw an analogy with unjust laws concerning homosexuality. In many jurisdictions from the 1950s to 1970s homosexuality was criminalised, and individuals who broke such laws could be legally compelled to receive medical interventions or could be offered such interventions as an alternative to incarceration. In the words of Hubner and White, 'A homosexual is not relieved from any individual suffering when "treated" [for] his condition, but just from the (unjustified) social sanctions that he has to face. Analogously, a psychopath is not freed from any personal distress when treated for her disposition, but just from the (well-founded) social penalties that she is subjected to'.[41] Furthermore, Hubner and White argue that even if psychopaths agree to treatment, their agreement cannot truly be voluntary if they are motivated by external pressure (e.g., the chance of early release from prison if the treatment is successful). Again, returning to the analogy with homosexuality, they argue

[40] D Hubner and L White, 'Neurosurgery for Psychopaths? An Ethical Analysis' (2016) 7(3) *American Journal of Bioethics Neuroscience* 140.

[41] Ibid.

that if someone sought 'treatment' for his/her homosexuality, their consent would probably be invalid, as a desire for this treatment could only be motivated by external pressure. They argue that the nature of psychopathy means that psychopaths' motivation to seek treatment cannot arise from psychopaths' internal motivations because they do not find their condition problematic.

In response, I will argue, first, that psychopaths do suffer as a *direct* consequence of their condition (rather than just as a result of punishment). Second, I will argue that being motivated to accept treatment by the desire for early release from detention does not necessarily undermine the voluntariness of the treatment.

As I discussed earlier, when describing the nature of psychopathy, psychopaths constantly engage in self-defeating behaviour, which results in suffering, e.g., feelings of frustration. This self-defeating behaviour is a direct result of aspects of their condition, such as impulsivity and their problems with mental time travel. Psychopathy also results in a range of physical health problems and early death. The poor decision making that characterises the condition means that psychopaths typically form unhealthy habits, are prone to substance abuse, frequently get injured as a result of physical fights and are often killed or injured in car accidents, due to reckless driving.[42]

Hubner and White's claim that psychopaths do not find their condition problematic is misleading. They find the direct consequences of their condition problematic, but they fail to understand that their condition is the source of their suffering. Psychopaths typically lack insight into their condition and instead blame others for their problems. However, it is possible that talking therapy might enable psychopaths to gain at least a partial insight into the connection between their condition and some of their problems. This might provide some 'internal' motivation to seek further treatments, possibly including biomedical interventions. (Of course, the psychopath's participation in the initial talking therapy might be motivated by 'external' incentives. However, the ethical preconditions for participating in safe, non-invasive interventions are arguably less stringent than for biomedical interventions.)

Furthermore, even if psychopaths were motivated by external factors, e.g., the desire for early release from prison, this would not necessarily render their consent to biomedical treatments involuntary. To illustrate

[42] Olli Vaurio et al., 'Psychopathy and Mortality' (2017) *Journal of Forensic Sciences* (Online First) Doi: 10.1111/1556-4029.13566.

this point, I will propose an alternative analogy – which I will call the 'quarantine example'. The treatment of psychopathy is more similar to the quarantine example than to Hubner and White's favoured analogy with homosexuality.

> *The Quarantine Example:* Imagine that Mary carries a disease that causes her absolutely no harm or discomfort. However, this disease is extremely infectious and would cause great harm or death to many of those who would become infected.

It seems intuitively plausible that the authorities would have the right to quarantine Mary, if this were necessary to prevent her from spreading the disease. Now imagine that a treatment is discovered that would render Mary noninfectious. If Mary understands the potential risk of side effects and would prefer to take the drug rather than to spend a considerable period in quarantine, it is not obviously unethical to give her the treatment. Mary's situation is completely different from the situation of the person convicted under unjust antihomosexuality laws and offered treatment in exchange for early release. One of the key differences is that the state clearly has no legitimate reason to detain or treat the homosexual person. In contrast, the state is entitled to detain Mary in order to protect others. In this respect, Mary's situation is similar to a dangerous psychopath's. The state is entitled to detain those psychopaths who have demonstrated by their behaviour that they pose a serious threat to other. (I will not attempt to settle the question of whether this should be criminal or civil detention.) If Mary can validly consent to receive treatment rather than quarantine, then it seems plausible that the psychopath can validly consent to treatment rather than incarceration or civil detention.

However, Nicole Vincent has raised an objection against using neurointerventions to alter psychopathic traits, which might reveal a difference between Mary and the psychopath. She points out that neither philosophers nor psychiatrists have devised a clear way of determining whether psychopaths' 'symptoms' are *incapacities* or *character flaws.*[43] Psychopaths can be described as 'cold, shallow superficial . . . deceitful . . . impulsive . . . careless [and] lacking forethought'. Vincent notes that this sounds more like a description of condemnable character traits rather than the description of a disease. She argues that altering character traits

[43] N Vincent, 'Assessment and Modification of Free Will through Scientific Techniques: Two challenges' in W Glannon (ed.), *Free Will and the Brain* (Cambridge University Press 2015).

is much more morally problematic than treating a disease or even restoring/creating/enhancing mental capacities. If Vincent is right, then it may be ethical for Mary to be offered the choice between quarantine or treatment for her infectious disease, but it may not be ethical to offer a psychopath treatment for his 'character flaws'. Acknowledging that the state is entitled to detain psychopaths to prevent them from harming others does not give the state permission to 'deny them the right to think what they want to think and to be who they want to be'.[44]

I agree with Vincent that attempting to fundamentally remould someone's character through neurointerventions would be morally problematic. If the recipient only agrees due to external pressure, this 'agreement' may not be enough to make the practice acceptable. Respect for agency requires that attempts to alter an offender's character involve engaging with their 'capacity for reason and reflection'.[45] I have argued elsewhere that simply trying to reprogramme an offender's character, without providing the offender with good reasons for changing, risks objectifying the offender, even if the offender does consent. Such interventions also arguably objectify those offenders who refuse treatment, by implying that they stand in need of such radical reprogramming and lack an important aspect of agency – the ability to mend one's own character flaws.[46]

Nevertheless, this objection need not preclude the use of all neurointerventions to treat psychopathy. Respect for agency can be shown by combining neurointerventions with moral dialogue and talking therapy and by ensuring that neurointerventions aim to facilitate the exercise of agency, rather than fundamentally remoulding an offender's character. Although, Vincent is correct that it is impossible to draw a sharp line between enhancing mental capacities and fundamentally remoulding a person's character, this does not mean that there is no meaningful distinction between these two things. An attempt to directly rewire a psychopath's brain to make him warm and highly compassionate seems like an unacceptable remoulding of character. However, a more subtle increase in the ability to empathise (which he can choose not to exercise) seems more like a capacity enhancement. Interventions that increase cognitive skills, such as increasing the ability to focus on relevant considerations, also seem more like capacity enhancements rather than remoulding someone's character (provided that the individual can choose not to focus, if he wishes).

[44] Ibid., p 180. [45] Ibid., p. 179.

[46] Elizabeth Shaw, 'Direct Brain Interventions and Responsibility Enhancement' (2014) *Criminal Law and Philosophy*, 8(1), 1–20.

Some Ethical Preconditions for the Treatment of Psychopathic Offenders

This section will propose a number of preconditions that should be satisfied before neurointerventions can permissibly be given to psychopathic offenders. The proposed approach attempts to strike a balance between respecting psychopaths' autonomy and promoting their welfare. I will assume in this section that the individual being offered the treatment would otherwise be rightfully detained for public protection. It is beyond the scope of this chapter to discuss whether this should be civil detention or imprisonment. I will also assume that the purpose of giving the treatment is rehabilitation. I will not consider the possibility of imposing neurointerventions as a form of retributive punishment.[47]

1. *The characteristic targeted by the treatment must be one that the state is entitled to alter.*

   As noted earlier, the state is not entitled to alter characteristics that are in no way connected with a risk of harm (e.g., sexual orientation), and the state is not entitled to remould an offender's character. However, interventions that increase the offender's capacity to obey the law (and which allow him to choose whether to exercise this capacity) seem more acceptable.[48]
2. *Neurointerventions should only be given with the offender's informed consent*

   The individual must be given full information about the treatment in terms that are as accessible as possible. Due to the nature of psychopathy, it may be impossible for the individual to have a full appreciation of what the effects of the neurointervention may be like. For instance, a person with little or no empathy may find it hard to imagine what an increase in empathy might be like. However, many medical interventions pose similar problems and yet are not unethical (e.g., interventions to restore sight or hearing to someone whose senses of sight and hearing have been impaired from birth). Ideally, the treatment should be given incrementally and should be reversible, so that the individual can withdraw if he/she objects to the treatment's effects.

[47] This issue is discussed in E. Shaw Retributivism and the biomedical moral enhancement of offenders' in Michael Hauskeller (ed.) *Moral Enhancement* (Cambridge University Press forthcoming).

[48] L. Bomann-Larsen, 'Voluntary Rehabilitation? On Neurotechnological Behavioural Treatment, Valid Consent and (In)appropriate Offers' (2013) 6(1) *Neuroethics* 65.

The state must not coerce the offender into accepting the intervention. In other words, this precludes putting *illegitimate* pressure on the offender to accept the treatment. Offering to reduce an offender's sentence if the treatment is successful does not necessarily involve illegitimate pressure. The state has the right to protect members of the community from psychopaths' harmful behaviour, but if the psychopath ceases to be harmful due to successful treatment, it seems reasonable to release him. However, it is not obvious that the state has the right to pressure the offender into choosing the treatment *rather than* detention. Therefore the state should be neutral as to whether the offender chooses treatment or detention. It should not *intend*, when making the offer that the offender chooses treatment.[49]

3. *Neurointerventions that are significantly worse than detention should not be offered.*

Given that the coercive environment of the criminal justice system means that the psychopathic offender is in a vulnerable position, some writers have warned that certain offers of treatment may be exploitative.[50] Focquaert suggests that this consideration might rule out invasive neurointerventions with potentially moderate to severe side effects.[51] I agree that such interventions will often be unethical. However, this will depend on what the alternative is. It is possible to imagine that a very lengthy period of detention (which may be entirely justified, given the individual's dangerousness) may objectively set back the offender's interests to the same degree or to a greater degree than a neurointervention with moderate side effects. If the offender prefers to accept the neurointervention, then is not obvious that the offender should be denied such treatment. However, it would be unethical for the state knowingly to offer a treatment that is clearly worse for the recipient than continued detention.

4. *The state should respect the offender's agency*

Neurointerventions should be combined with moral dialogue about the reasons why they should change their behaviour and should aim to facilitate the exercise of agency. 'Active' interventions that produce their

[49] E Shaw, 'Offering Surgical Castration to Sex Offenders: The Significance of the State's intentions' (2014) 40(9) *Journal of Medical Ethics* 594.

[50] J McMillan, 'The Kindest Cut? Surgical Castration, Sex Offenders and Coercive Offers' (2013) *Journal of Medical Ethics* (Online First) *doi:10.1136/medethics-2012-101030;* B Foddy and J Savulescu 'Addiction and Autonomy: Can Addicted People Consent to the Prescription of Their Drug of Addiction?' (2006) 20 *Bioethics* 1.

[51] F Focquaert (2014).

effects via the agent's active involvement (e.g., neurofeedback) are preferable to 'passive' interventions that produce their effects irrespective of the agent's will (e.g., drugs or neurosurgery).[52]

Conclusion

This chapter discussed the possibility that reasons of principle (as opposed to empirical reasons) might mean that it would be futile to attempt to alter psychopaths' problematic character traits or behaviour using biomedical means. It also considered the possibility that there is something about the nature of psychopathy, which means that it could *never* be ethically permissible to offer offenders of this sort any treatments that potentially carry significant risks. Both possibilities were rejected. The chapter then proposed a number of preconditions that should be satisfied before neurointerventions could permissibly be given to psychopathic offenders. Given the slow progress in developing treatments for psychopathy and the morally repellent nature of many psychopaths' behaviour, there is a strong temptation to consider the search for treatments to be futile and to opt instead for a purely punitive approach. However, as I have argued, this temptation should be resisted.

52 F Focquaert and M Schermer, 'Moral Enhancement: Do Means Matter Morally?' (2015) 8(2) *Neuroethics* 139.

Appendix

Three examples of expert reliability issues, each involving neuropathology arose in the context of alleged baby shaking. The names of the subjects of criticism, although in the published judgments, have been removed to avoid distraction. Examples 1 and 2 are from the judgment of Lord Justice Moses in two of the joined appeals in *Keran Henderson and others* [2010] EWCA Crim 1269. Example 3 is from the judgment of Mrs Justice King in the Family Division in *A Local Authority v S* [2009] EWHC 2115.

EXAMPLE 1

88. . . . Dr S's stance, in oral evidence before us, casts significant doubt upon the reliability of the rest of her evidence and her approach to this case. It demonstrates, to our satisfaction, that she was prepared to maintain an unsubstantiated and insupportable theory in an attempt to bolster this appeal.

Example 2. (in which the expert, Dr L, had been called to challenge the prosecution's case that it was possible to distinguish axonal damage caused by trauma from axonal damage caused by hypoxia, a technique using Beta APP testing developed by Dr Al-Sarraj at King's College, London; none of the matters that attracted the Court of Appeal's criticisms were disclosed in Dr L's reports. To understand the context, see paras. 48 et seq of the judgment).

EXAMPLE 2

55. He also commented that beta app staining 'was not available when he was doing this kind of work' . . . .

61. In our judgment, Dr L's experience was more historic and far more limited than that of Dr Al-Sarraj. He has not conducted autopsies or given evidence in cases involved with baby-shaking for many years. His knowledge emerges from studies of the literature which excluded the important

(CONT.)

material contained in Geddes I and II and Reichard. Until his meeting with Dr Al-Sarraj, he had never appreciated the significance of the signs of axonal damage in the cortico-spinal tract. His lack of experience and his lack of appreciation of the importance of this point leads us to the conclusion that his evidence was fundamentally flawed. His insistence on asserting that it was not possible to attribute significance to that axonal damage flies in the face of the references in both Geddes I, II and Reichard and the evidence which flows from Dr Al-Sarraj's up-to-date experience. We reject his evidence insofar as it fails to attribute significance to that which the beta app staining revealed.

63. . . . the willingness of Dr L to advance propositions which he subsequently had to withdraw . . . coupled with his lack of up-to-date experience, severely damaged and undermined the effect of his evidence. We would have had considerable doubts as to whether he was properly qualified to give evidence designed to refute the evidence given by [prosecution experts] at trial.

Example 3 – Mrs Justice King in *A Local Authority v S* [2009] EWHC 2115:

EXAMPLE 3

245 *Any Court dealing with these cases on a regular basis will be aware that Dr. S is a strong believer that many cases hitherto regarded as SBS [Shaken Baby Syndrome] have in fact been precipitated by a choke and that there has thereafter been a cascade of events, including hypoxia, which has been responsible for the subdural haemorrhages and retinal haemorrhages.*

246 *Dr S's view is a legitimate one and an appropriate line of research. All agree that much remains unknown about SBS and the triad. It is essential, however, that Dr. S and others engaged on such research avoid becoming a zealots with the consequence that scientific rigour is lost or sacrificed.*

247 *These Courts rely on the professionalism and rigor of the experts who come before them. That means not only drawing the Court's attention to research that is contrary to their view, but that the experts are rigorous in the use they make of research papers. Dr. S's suggestion that there were 'well documented cases of choking, leading to subdural haemorrhages and retinal haemorrhages' by reference to the two articles analysed above is disquieting. It should be borne in mind that Dr. S relied upon these so*

(CONT.)

called 'well documented cases' in support of her hypothesis as to the cause of Z's death; the subject of this enquiry. She should have, at the very least, drawn the court's attention to the fact that the cases in question were cases where the child had been shaken and that in one of them had been regarded as a case of non accidental injury resulting in the child being taken into care as a result of the incident.

248 Dr. Bonshek referred to Dr. S's use of these articles as being 'disingenuous'. I feel driven, with regret, to agree.

(iii) Factual accuracy
259 These cases are difficult, distressing and complex. I understand that Dr. S and Dr. C have strong views about Shaken Baby Syndrome and the triad. They are each to be applauded for their dedication to the cause of research in this controversial area and it is to their individual credit that they are undeterred by being, as they see it, voices in the wilderness.

260 However, it is of the utmost importance that all experts, whether mainstream or not, read all the papers and where they have to rely on raw data that they check its veracity and accuracy in the medical notes. A trial is first and foremost, a forensic exercise and fairness to the parties demands, as a basic premise, that the experts will be accurate in their use of the source material.

284 I do not doubt the commitment of Dr. S and Dr. C to the advancement of the understanding of Shaken Baby Syndrome. As already indicated, I make no criticism and, indeed, it would be wrong to do so, of the fact that neither of them hold mainstream views. There is a significant fundamental difference between academic theories and hypotheses, on the one hand, and the rigorous forensic analysis which is required in care proceedings, on the other. In care proceedings the parents of the children concerned face allegations of the most serious type and they are therefore entitled to expect the experts commissioned to report to the court to be meticulous in both their analysis of the data and in their presentation to the court of their expert forensic opinion. In addition uniquely in this division, the Court is concerned with the future protection and safety of the children of those parents.

285 Dr. S and Dr. C, I find with regret, have each fallen into that category of expert identified by Butler-Sloss P. in Re LU & LB, namely the expert who has developed a scientific prejudice.

BIBLIOGRAPHY

Abend, Gabriel 'What Are Neural Correlates Neural Correlates of?' (2016) *BioSocieties* 12(3), 1–24.

Aguilar, Jesus, and Buckareff, Andrei (eds.) *Causing Human Actions: New Perspectives on the Causal Theory of Action* (Cambridge, MA: MIT Press, 2010), pp. 27–43.

American Psychiatric Association (2013) *Diagnostic and Statistical Manual of Mental Disorders, 5th ed. (DSM 5)* (Washington, DC: American Psychiatric Association).

Anderson, E. 'What Is the Point of Equality?' (1999) *Ethics* 2, 287–337.

Anon 'This Machine Records All Your Thoughts' (1919) The Syracuse Herald Available at: www.paleo-future.blogspot.nl/2007/05/this-machine-records-all-your-thoughts.html.

Anscombe, G. E. M. *Intention* (Oxford: Blackwell, 1957).

Anscombe, G. E. M., and Geach, P. *Descartes: Philosophical Writings* (Indianapolis, IN: Bobbs-Merrill Company, 1954).

Arico, Adam J., and Fallis, Don. 'Lies, Damned Lies, and Statistics: An Empirical Investigation of the Concept of Lying' (2013) *Philosophical Psychology* 26 790–816.

Asp, P. 'Previous Convictions and Proportionate Punishment under Swedish Law' in J. Roberts and A. Von Hirsch (eds.) *Previous Convictions at Sentencing* (Oxford: Hart Publishing, 2010), pp. 207–26.

Audi, R. 'Acting for Reasons' (1986) *The Philosophical Review* 95, 511–46.

Reasons, Rights, and Values (Cambridge: Cambridge University Press, 2015).

Austin, John. *Lectures on Jurisprudence, Or, The Philosophy of Positive Law* (London: J. Murray, 1869), Lecture XVIII.

Bargh J. 'Our Unconscious Mind' (2013) *Scientific American* 310(1).

Baron-Cohen, S. *Mindblindness: An Essay on Autism and Theory of Mind (Learning, Development and Conceptual Change)* (Cambridge, MA: MIT Press, 1995).

Bechara, Antoine, Damasio, Antonio, Damasio, Hanna, and Anderson, Steven. 'Insensitivity to Future Consequences Following Damage to Human Prefrontal Cortex' (1994) *Cognition* 50, 7–15.

Bechtel, William. *Mental Mechanisms: Philosophical Perspectives on Cognitive Neuroscience* (Mahwah, NJ: Erlbaum, 2007).

Bennett M. R., and Hacker, P. M. S. *Philosophical Foundations of Neuroscience* (Malden, MA: Blackwell, 2003).

Ben-Shakhar, G. 'Effects of Mental Countermeasures on Psychophysiological Detection in the Guilty Knowledge Test' (1991) *International Journal of Psychophysiology* 11(2), 99–108.

'A Critical Review of the Control Questions Test' in M. Kleiner (ed.) *Handbook of Polygraph Testing* (London: Academic Press, 2002).

Ben-Shakhar, G., and Dolev, K. 'Psychophysiological Detection through the Guilty Knowledge Technique: Effects of Mental Countermeasures' (1996) *Journal of Applied Psychology* 81(3), 273–81.

Berker, Selim 'The Unity of Grounding,' working paper, 2015.

Birbaumer, N., Veit, R., Lotze, M., et al. 'Deficient Fear Conditioning in Psychopathy: A Functional Magnetic Resonance Imaging Study' (2005) *Archives of General Psychiatry* 62(7), 799–805.

Blair, J., Mitchell, D., and Blair, K. *The Psychopath: Emotion and the Brain* (Oxford: Blackwell, 2005).

Blair, R. J. R., Colledge, E., Murray, L., and Mitchell, D. G. V. 'A Selective Impairment in the Processing of Sad and Fearful Expressions in Children with Psychopathic Tendencies' (2001) *Journal of Abnormal Child Psychology* 29(6), 491–8.

Blair, R. J. R., Mitchell, D. G. V., Leonard, A., Budhani, S., Peschardt, K. S., and Newman, C. 'Passive Avoidance Learning in Individuals with Psychopathy: Modulation by Reward but Not by Punishment' (2004) *Personality and Individual Differences* 37(6), 1179–92.

Blair, R. J. R., Peschardt, K. S., Budhani, S., Mitchell, D. G. V., and Pine, D. S. 'The Development of Psychopathy' (2006) *Journal of Child Psychology and Psychiatry* 47(3–4), 262–76.

Blodwell, J. 'Physicalism and Emergence' (2011) *Journal of Consciousness Studies* 18, 6–31.

Blum, G. S. *Psychodynamics: The Science of Unconscious Mental Forces.* (Belmont, CA: Wadsworth Publishing, 1968).

Blume, C., del Giudice, R., Wislowska, M., Lechinger, J., and Schabus, M. 'Across the Consciousness Continuum – from Unresponsive Wakefulness to Sleep' (2015) *Frontiers in Human Neuroscience* 9, 1–14.

Bonta, J., and Andrews, D. *The Psychology of Criminal Conduct* (London: Routledge, 2017).

Bostrom, N., and Ord, T. 'The Reversal Test: Eliminating Status Quo Bias in Applied Ethical Reasoning' (2006) *Ethics* 116, 656–79.

Brandom, Robert. *Making It Explicit* (Cambridge, MA: Harvard University Press, 1994).

Bratman, M. *Faces of Intention: Selected Essays on Intention and Agency* (Cambridge: Cambridge University Press, 1999).

Brinkmann, Sevend 'Mental Life in the Space of Reasons' (2006) *Journal for the Theory of Social Behaviour* 36, 1.

Brooks, David Op-Ed, *New York Times*, June 17, 2013.

Brown, Teneille, and Murphy, Emily 'Through a Scanner Darkly: Functional Neuroimaging as Evidence of a Criminal Defendant's Past Mental States' (2012) *Stanford Law Review* 62, 1119–208.

Bublitz, J.-C. 'My Mind Is Mine?! Cognitive Liberty as a Legal Concept' in E. Hildt and A. Francke (eds.) *Cognitive Enhancement* (New York: Springer, 2013).

Buckholtz Joshua W., and Faigman, David L. 'Promises, Promises for Neuroscience and Law' (2014) *Current Biology* 24(3), 437–40.

Burgess, A. *A Clockwork Orange* (London: Heinemann, 1962).

Burns Jeffrey, and Swerdlow, Russell 'Right Orbitofrontal Tumor with Pedophilia Symptom and Constructional Apraxis Sign' (2003) *Archives of General Neurology* 60, 437–40.

Byrne, A. J., Sellen, A. J., and Jones, J. G. 'Errors on Anaesthetic Record Charts as a Measure of Anaesthetic Performance during Simulated Critical Incidents' (1998) *British Journal of Anaesthesiology* 80(1), 58–62.

Cacioppo, John, and Moore, Michael 'Decisions and Intentions Working Group Paper No. 1,' MacArthur Foundation Project on Law and Neuroscience, 2008.

Carson, Thomas L. 'The Definition of Lying' (2006) *Nous* 40, 284.

Caruso, G. *Free Will and Consciousness: A Determinist Account of the Illusion of Free Will* (Langham, MD: Lexington Books, 2012).

'Free Will Eliminativism: Reference, Error, and Phenomenology' (2015) *Philosophical Studies* 172(10), 2823–33.

'Free Will Skepticism and Criminal Behavior: A Public Health-Quarantine Model' (2016) *Southwest Philosophy Review* 32(1), 24–48.

The Case of Edmund Peacham (1615) 2 How. St. Tr. 869.

Case, S. and Haines, K. *Understanding Youth Offending* (Cullompton: Willan, 2009).

Chabris C. F., and Simons D. J. *The Invisible Gorilla: Thinking Clearly in a World of Illusions* (New York: Harper Collins Publishers, 2010).

Chalmers, D. 'Facing up to the Problem of Consciousness' (1995) 2 *Journal of Consciousness Studies* 200–19.

Changeux, Jean-Pierre. *The Good, the True and the Beautiful: A Neuronal Approach* (New Haven, CT: Yale University Press, 2012).

Chisholm, Roderick ed., *Realism and the Background of Phenomenology* (Glencoe, IL: Free Press, 1960).

Churchland, P. S. *Neurophilosophy: Toward a Unified Science of the Mind-Brain*, (Cambridge, MA: MIT Press, 1986).

'Moral Decision-Making and the Brain,' in Judy Illes (ed.), *Neuroethics: Defining the Issues in Theory, Practice and Policy* (Oxford: Oxford University Press, 2005),

'Of Brains & Minds: An Exchange' (June 19, 2014) *New York Review of Books*, at www.nybooks.com/articles/2014/06/19/brains-and-minds-exchange/.

Churchland, Paul 'Eliminative Materialism and the Propositional Attitudes,' (1981) *Journal of Philosophy* 78, 67–90.

Churchland, P., and Churchland, P. S. *On the Contrary: Critical Essays, 1987–1997* (Cambridge, MA: MIT Press, 1999).

Cleckley, Henry *The Mask of Sanity*. 5th ed. (St. Louis, MO: Mosby, 1988).

Copas, J. B. *Statistical Analysis for a Risk of Reconviction Predictor* (1992) Report to the Home Office. University of Warwick: unpublished.

Corcoran, K. J. 'The Trouble with Searle's Biological Naturalism.' (2001) *Erkentnis* 55 (3).

Cornell, D. G., Warren, J., Hawk, G., Stafford, E., Oram, G., and Pine, D. 'Psychopathy in Instrumental and Reactive Violent Offenders' (1996) *Journal of Consulting and Clinical Psychology*, 64(4), 783–90

Cornet, L., Kogel, C., Nijman, H., Raine, A., and Laan, P. 'Neurobiological Changes after Intervention in Individuals with Antisocial Behaviour: A Literature Review' (2015) *Criminal Behaviour and Mental Health* 25(1), 10–27.

Covey, R. 'Signaling and Plea Bargaining's Innocence Problem' (2009). *Washington and Lee Law Review* 66, 73–130.

Cowen, R. et al., 'Assessing Pain Objectively: The Use of Physiological Markers' (2015) *Anaesthesia* 70(7), 828–47.

Crane, T. 'Mental Causation' in L. Nadel (ed.), *Encyclopedia of Cognitive Science* (New York: Wiley, 1995).

Craver, Carl. *Explaining the Brain: Mechanisms and the Mosaic Unity of Neuroscience* (Oxford: Oxford University Press, 2007).

Crick, Francis. *The Astonishing Hypothesis: The Scientific Search for the Soul.* (London: Touchstone Books, 1995).

Crick, N. R., and Dodge, K. A. 'Social Information-Processing Mechanisms in Reactive and Proactive Aggression' (1996) *Child Development* 67(3), 993–1002.

Criminal Justice Act. 2003.

Crockett, M et al., 'Serotonin Selectively Influences Moral Judgment and Behavior through Effects on Harm Aversion' (2010) *Psychological and Cognitive Sciences* 107 (40), 17433.

Cui, Qian et al. 'Detection of Deception Based on fMRI Activation Patterns Underlying the Production of a Deceptive Response and Receiving Feedback about the Success of the Deception after a Mock Murder Crime' (2013) *Social Cognitive and Affective Neuroscience* 9(10), 1472–81.

Cummins, R., *Meaning and Mental Representation* (Cambridge, MA: MIT Press, 1989).

Dadds, M., Moul, C., Cauchi, A., Dobson-Stone, C., Hawes, D., Brennan J., and Ebstein, R. 'Methylation of the Oxytocin Receptor Gene and Oxytocin Blood Levels in the Development of Psychopathy' (2014) *Development and Psychopathology* 26(1), 33.

Danaher, J. 'Scientific Evidence and the Criminal Law: Lessons from Brain-Based Lie Detection' (2010) *Judicial Studies Institute Journal* 10(1), 94.

'The Future of Brain-Based Lie Detection and the Admissibility of Scientific Evidence' (2011) *Irish Criminal Law Journal* 21(4), 67–76.

'The Comparative Advantages of Brain-Based Lie Detection: The P300 Concealed Information Test and Pre-Trial Bargaining' (2015a) *International Journal of Evidence and Proof* 19(1), 52–66.

'Responsible Innovation in Social Epistemic Systems: The P300 Memory Detection Test and the Legal Trial' in B.-J. Koops et al. (eds.) *Responsible Innovation Volume II: Concepts, Approaches, Applications* (Dordrecht: Springer, 2015b).

Daubert v. Merrell Dow Pharm., Inc. 509 U.S. 579 (1993).

Davidson, Donald. 'Actions, Reasons, and Causes' (1963) *The Journal of Philosophy* (60) 685–700.

Essays on Actions and Events (Oxford: Clarendon Press, 1980).

Davis, N. and Koningsbruggen, M. 'Non-Invasive' Brain Stimulation Is Not Non-Invasive' (2013) *Frontiers in Systems Neuroscience* 7(76), 1–4.

Davis, N., Tomlinson, S. P., and Morgan, H. M. 'The Role of Beta-Frequency Oscillations in Motor Control' (2012) *Journal of Neuroscience* (32), 403–4.

De Ganck, J., and Vanheule, S. 'Bad Boys Don't Cry': A Thematic Analysis of Interpersonal Dynamics in Interview Narratives of Young Offenders with Psychopathic Traits' (2015) *Frontiers in Psychology* (6) 960.

Dennett, Daniel. 'Skinner Skinned' in *Brainstorms* (Putney, VT: Bradford Books, 1978).

Denson, E. F. et al. 'The Angry Brain: Neural Correlates of Anger, Angry Rumination, and Aggressive Personality' (2009) *Journal of Cognitive Neuroscience* 21(4), 734–44.

De Ridder, D., Langguth, B., Plazier, M., and Menovsky, T. 'Moral Dysfunction: Theoretical Model and Potential Neurosurgical Treatments' in J. Verplaetse, S. Vanneste, J. Braeckman, and J. Schrijver (eds.) *The Moral Brain: Essays on the Evolutionary and Neuroscientific Aspects of Morality* (New York: Springer, 2009), pp. 155–84.

Deutsch, D. *The Fabric of Reality* (London: Penguin Books, 1997).

Dodge, K. A., and Coie, J. D. 'Social-Information-Processing Factors in Reactive and Proactive Aggression in Children's Peer Groups' (1987) *Journal of Personality and Social Psychology*, 53(6), 1146–58.

Dolan B., and Coid, J. 'Psychopathic and Antisocial Personality Disorders: Treatment and Research Issues' (Gaskell/Royal College of Psychiatrists 1993).

Donald B., and Bakies, E. 'A Glimpse Inside the Brain's Black Box: Understanding the Role of Neuroscience in Criminal Sentencing' (2016) *Fordham Law Review* 85, 481–502.

Donnelly-Lazarov, *A Philosophy of Criminal Attempts* (Cambridge: Cambridge University Press, 2015).

'Intention in Criminal Law: The Challenge from Non-Observational Knowledge,' (2017) *Ratio Juris*.

'What it means to be responsible for action: Mind and Movement in Criminal Law' (2017) *Journal of Law Information and Science* 25(3).

'Intention in Criminal Law: the Challenge from Non-Observational Knowledge' (2017) *Ratio Juris* 30(4), 451–70.

Douglas, T. 'Moral Enhancement' (2008) *Journal of Applied Philosophy* 25 (3), 228–45.

Dretske, F. *Explaining Behavior: Reasons in a World of Causes* (Cambridge, MA: MIT Press, 1998).

Dunne v Kier Group [2000] C.L.Y. 326.

Dworkin, R. 'What Is Equality? Part 1: Equality of Welfare' (1981a) *Philosophy & Public Affairs*, 10(3), 185–246.

'What Is Equality? Part 2: Equality of Resources' (1981b) *Philosophy & Public Affairs*, 10(4), 283–345.

Eccles, John 'The Initiation of Voluntary Movements by the Supplementary Motor Area' (1982) *Archives of Psychiatry and Neurological Sciences* 231, 423–39.

How the Self Controls Its Brain (Berlin: Springer Science, 1994).

Egan, G. et al. 'Neural Correlates of the Consciousness of the Emergence of Thirst' (2003) *Proceedings of the National Academy of Sciences of the United States of America* 100(25), 15241–6.

Fagan, T., Hirstein W., and Sifferd K. 'Innocent Minds: Child Soldiers, Executive Functions, and Culpability' (2016) *International Criminal Law Review* 16(2), 258–86.

Fallis, Don. 'What Is Lying?' (2009) *Journal of Philosophy* 106(1), 29–56.

'Lying as a Violation of Grice's First Maxim of Quality' (2012) *Dialectica* 66(4), 563–81.

Falvey, Kevin. 'Knowledge in Intention' (2000) *Philosophical Studies* 99, 21–44.

Farah, M., and Hook, C. 'The Seductive Allure of "Seductive Allure"' (2013) *Perspectives on Psychological Science* 8, 88–90.

Farah Martha J. et al. 'Functional MRI-Based Lie Detection: Scientific and Societal Challenges' (2014) *Nature Reviews Neuroscience* 15, 123–31.

Farrall, S., Hunter, B., Sharpe, G., and Calverley, A. *Criminal Careers in Transition: The Social Context of Desistance from Crime* (Oxford: Oxford University Press, 2014).

Farwell, L. 'Brain Fingerprinting: A Comprehensive Tutorial Review of Detection of Concealed Information with Event-Related Brain Potentials' (2012) *Cognitive Neurodynamics* 6, 115–54.

Farwell, L., and Donchin, E. 'The Truth Will Out: Interrogative Polygraphy ('Lie Detection') with Event Related Potentials' (1991) *Psychophysiology* 28(5), 531–47.

Farwell, L., Richardson, D., and Richardson, G. 'Brain Fingerprinting. Field Studies Comparing P300-MERMER and P300 Brainwave Responses in the Detection of Concealed Information' (2013) *Cognitive Neurodynamics* 7(4), 263–99.

Farwell, L., and Smith S. S. 'Using Brain MERMER Testing to Detect Knowledge Despite Efforts to Conceal' (2001) *Journal of Forensic Science* 46, 135–43.

Fecteau, S., Knoch, D., Fregni, F., Sultani, N., Boggio, P., and Pascual-Leone, A. 'Diminishing Risk-Taking Behavior by Modulating Activity in the Prefrontal Cortex: A Direct Current Stimulation Study' (2007) *The Journal of Neuroscience* 27 (46), 12500–5.

Feigl, Herbert et al., eds. *Minnesota Studies in the Philosophy of Science*, Vol. 2 (Minneapolis: University of Minnesota Press, 1958).

Ferari, P. et al., 'Escalated Aggressive Behavior: Dopamine, Serotonin and GABA' (2005) *European Journal of Pharmacology* 526(1–3), 51–64.

Field v Leeds City Council [2001] C.P.L.R. 129.

Filevich, E. et al. 'Brain Correlates of Subjective Freedom of Choice' (2013) *Consciousness and cognition* 22(4), 1271–84.

Fine, Kit. 'Guide to Ground' in F. Correia and B. Schneider (eds.) *Metaphysical Grounding* (Cambridge: Cambridge University Press, 2012), 37–80.

Fischer, J. M., and Ravizza M. *Responsibility and Control: A Theory of Moral Responsibility* (Cambridge: Cambridge University Press, 1999).

Fleischman, John. *Phineas Gage: A Gruesome but True Story About Brain Science* (New York: Houghton, Mifflin, 2002).

Fletcher, George. *Rethinking Criminal Law* (Boston: Little, Brown, 1978).

Flew, A. *Crime or Disease?* (London: Macmillan, 1973).

Focquaert, F. 'Mandatory Neurotechnological Treatment: Ethical Issues' (2014) *Theoretical Medicine and Bioethics* 35(1), 59–72.

Focquaert F., and Schermer, M. 'Moral Enhancement: Do Means Matter Morally?' (2015) *Neuroethics* 8(2), 139–51.

Foddy B., and Savulescu J. 'Addiction and Autonomy: Can Addicted People Consent to the Prescription of Their Drug of Addiction?' (2006) *Bioethics* 20(1), 1–15.

Fodor, Jerry. 'Special Sciences, or the Disunity of Science as a Working Hypothesis' (1974) *Synthese* 28, 97–115.

'The Language of Thought' (1975) *The Philosophical Review* 20(1), 59–63.

'Propositional Attitudes' (1978) *The Monist* 61(4), 501–24.

'Fodor's Guide to Mental Representation: The Intelligent Auntie's Vade-Mecum' (1985) *Mind* 373, 76–100.

Psychosemantics: The Problem of Meaning in The Philosophy of Mind (Cambridge, MA: MIT Press, 1987).

Fodor, J. A., and Pylyshyn, Z. W., 'Connectionism and Cognitive Architecture: A Critical Analysis' (1988) *Cognition* 28(1–2), 3–71.

Forth A et al. *The Psychopathy Checklist – Youth Version* (Toronto, Ontario: Multi-Health Systems, 2003).

Fox Dov, and Stein, Alex 'Dualism and Doctrine' in Dennis Patterson and Michael Pardo (eds.) *Philosophical Foundations of Law and Neuroscience* (Oxford: Oxford University Press, 2016), 126–35.

Frank R. *Passions within Reason: The Strategic Role of the Emotions* (New York: W. W. Norton, 1988).

Frick, Paul quoted by Stenson, Jacqueline 'Destined as a Psychopath: Experts Seek Clues,' www.nbcnews.com/id/30267075/ns/health-mental_health/t/destined-psychopath-experts-seek-clues/#.WPKnao61sXo Last accessed 15/04/17.

Frick, P., and Hare, R. *The Anti-Social Process Screening Device* (Toronto, Ontario: Multi-Health Systems, 2001).

Fumerton, Richard *Knowledge, Thought, and the Case for Dualism* (Cambridge: Cambridge University Press, 2013).

Funkhouser, E. 'Multiple Realizability' (2007) 2 *Philosophy Compass* 303–15.

Furedy, J. J. 'The "Control" Question "Test" (CQT) Polygrapher's Dilemma: Logico-Ethical Considerations for Psychophysiological Practitioners and Researchers' (1993) *International Journal of Psychophysiology* 15(3), 263–7.

Furedy, J., and Heselgrave, R. 'The Validity of the Lie Detector: A Psychophysiological Perspective' (1988) *Criminal Justice and Behavior* 15, 219–46.

Gamer, M. 'Detecting of Deception and Concealed Information Using Neuroimaging Techniques' in B. Verschuere, G. Ben-Shakhar, and E. Meijer (eds.) *Memory Detection: Theory and Application of Concealed Information Test* (Cambridge: Cambridge University Press, 2011).

Geach, P. *Mental Acts* (London: Routledge and Kegan-Paul, 1959).

Gettier, Edmund L. 'Is Justified True Belief Knowledge?' (1963) *Analysis* 6, 121–23.

Gibson, J. J. *The Ecological Approach to Visual Perception* (Boston, MA: Houghton Mifflin, 1979).

Glannon, W. 'Intervening in the Psychopath's Brain' (2014) *Theoretical Medicine and Bioethics* 35(1), 43–57.

Glenn, A. L. 'The Other Allele: Exploring the Long Allele of the Serotonin Transporter Gene as a Potential Risk Factor for Psychopathy: A Review of the Parallels in Findings' (2011) *Neuroscience & Biobehavioral Reviews* 35(3), 612–20.

Godman, M. and Jefferson, A. 'On Blaming and Punishing Psychopaths' (2017) *Criminal Law and Philosophy* 11(1), 127–42.

Goldberg, Gary 'Supplementary Motor Area Structure and Function: Review and Hypotheses' (1985) *Behavioral and Brain Sciences* (8), 567–88.

Goodman, Nelson *Fact, Fiction, and Forecast*, 4th ed. (Indianapolis: Bobbs-Merrill, 1984).

Gopnick, A., and H. M. Wellman 'Why the Child's Theory of Mind Really Is a Theory' (1992) *Mind & Language* (7) 145–71.

Greely, Henry T. 'Mind Reading, Neuroscience, and the Law' in Stephen J. Morse and Adina L. Roskies (eds.) *A Primer on Criminal Law and Neuroscience* (New York: Oxford University Press, 2013).

Greene, Joshua *Moral Tribes: Emotion, Reason, and the Gap between Us and Them* (New York: The Penguin Press, 2013).

Greene, Joshua and Cohen, Jonathan 'For the Law, Neuroscience Changes Nothing and Everything' (2004) *Philosophical Transactions of the Royal Society B: Biological Sciences* 359(1451), 1781–5.

'For Law, Neuroscience Changes Nothing and Everything' in S. Zeki and O. Goodenough (eds.) *Law and the Brain* (Oxford: Oxford University Press, 2006), pp. 207–26.

Greene, Joshua, and Paxton, Joseph M. 'Patterns of Neural Activity Associated with Honest and Dishonest Moral Decisions' (2009) *Proceedings of the National Academy of Sciences* 106(6), 12506–11.

Gregory, S., Blair, R. J., Simmons, A., et al. 'Punishment and Psychopathy: A Case-Control Functional MRI Investigation of Reinforcement Learning in Violent Antisocial Personality Disordered Men' (2015) *The Lancet Psychiatry* 2(2), 153–60.

Grice, H. P. 'Method in Philosophical Psychology (From the Banal to the Bizarre)' (1975) *Proceedings and Addresses of the American Philosophical Association*, 48, 23–53.

Hacker, P. M. S. *Human Nature: The Categorical Framework* (Oxford: Oxford University Press, 2007).

Hare, R. *Without Conscience: The Disturbing World of the Psychopaths among Us* (New York: The Guilford Press, 1999).

The Hare Psychopathy Checklist Revised, 2nd ed. (Toronto: Multi-Health Systems, 2003).

Hare, R. D., and Neumann, C. N. 'The PCL-R Assessment of Psychopathy: Development, Structural Properties, and New Directions' in C. Patrick (ed.) *Handbook of Psychopathy* (New York: Guilford, 2006).

Hare, Todd Camerer, Colin and Rangel, Antonio 'Self-Control in Decision-Making Involves Modulation of the vmPFC System,' (2009) *Science*, 646.

Harman, G. *Thought* (Princeton, NJ: Princeton University Press, 1973).

Harris C., and Wolpert, D. 'Signal-dependent Noise Determines Motor Planning' (1998) *Nature* 394, 780–784.

Hauskeller Michael (ed.) *Moral Enhancement* (Cambridge: Cambridge University Press, forthcoming).

Haynes, J.-D. 'Brain Reading: Decoding Mental States from Brain Activity in Humans' in J. Illes and B. J. Sahakian (eds.) *Oxford Handbook of Neuroethics* (Oxford: Oxford University Press, 2012), pp. 1–8.

Haynes, John-Dylan Sakai, Katsuyuki Rees, et al. 'Reading Hidden Intentions in the Brain' (2007) *Current Biology* 17 (2007) 323–328.

Henderson and others [2010] EWCA Crim 1269, [2010] 2 Cr App R 24.

Hirstein, W., and Sifferd K. 'The Legal Self: Executive Processes and Legal Theory' (2011) *Consciousness and Cognition* 20(1), 156–71.

Hofstadter Douglas R., and Dennett Daniel C. *The Mind's I* (New York: Bantam Books, 1981).

Hollin, C., and Palmer, E. (eds.) *Offending Behaviour Programmes* (Chichester: Wiley, 2006).

Holmes, Oliver Wendell *The Common Law* (Boston: Little, Brown, 1881).

Hooker, Clifford 'Toward a General Theory of Reduction, Part I (1981) *Dialogue* 20(1), 38–59.

'Toward a General Theory of Reduction, Part II (1981) *Dialogue*, 20(2), 201–36.

'Toward a General Theory of Reduction, Part III (1981) *Dialogue* 20(3), 496–529.

Horgan, Terence 'Nonreductive Physicalism and the Explanatory Autonomy of Psychology' in S. Wagner and R. Warner (eds.) *Naturalism: A Critical Appraisal* (South Bend, IN: University of Notre Dame Press, 1993), pp, 295–320.

Horikawa, T. et al. 'Neural Decoding of Visual Imagery during Sleep' (2013) *Science (New York, N.Y.)* 639–42.

Hornsby, Jennifer 'Causality and the Mental' (2015) *Humana.Mente Journal of Philosophical Studies* 29, 125–40.

H (Stephen) v The Queen [2014] EWCA Crim 1555.

Hu, X., Hegeman, D., Landry, E., and Rosenfeld, J. P. 'Increasing the Number of Irrelevant Stimuli Increases Ability to Detect Countermeasures to the P300-based Complex Trial Protocol for Concealed Information Detection' (2011) *Psychophysiology* 49(1), 85–95.

Hu, X., Pornpattananangkul, N., and Rosenfeld, J. P. 'N200 and P300 as Orthogonal and Integrable Indicators of Distinct Awareness and Recognition Processes in Memory Detection' (2013) *Psychophysiology* 50(5), 454–64.

Hu, X., and Rosenfeld, J. P. 'Combining the P300-complex Trial-based Concealed Information Test and the Reaction Time- based Autobiographical Implicit Association Test in Concealed Memory Detection' (2012) *Psychophysiology* 49(8), 1090–100.

Hubner D., and White, L. 'Neurosurgery for Psychopaths? An Ethical Analysis' (2016) *American Journal of Bioethics Neuroscience* 7(3), 140–49.

Hughes, M., Dolan, M., and Stout, J. 'Decision-making in Psychopathy' (2015) *Psychiatry, Psychology and Law* 23(4), 521–37.

Huis in 't Veld, E. M. J., and de Gelder, B. 'From Personal Fear to Mass Panic: The Neurological Basis of Crowd Perception Huis in 't Veld' (2015) *Human Brain Mapping* 36(6), 2338–51.

Hulsman, R. L., Ros, W. J. G., Winnubst, J. A. M., and Bensing, J. M. 'Teaching Clinically Experienced Physicians Communication Skills: A Review of Evaluation Studies' (1999) *Medical Education* 33(9), 655–68.

Hursthouse, Rosalind 'Intention' (2000) *Royal Institute of Philosophy Supplement* 46, 83–105.

Huth, A. G. et al. 'Natural Speech Reveals the Semantic Maps That Tile Human Cerebral Cortex' (2016) *Nature* 532, 453–58.

Hyman, John *Action, Knowledge, and Will* (Oxford: Oxford University Press, 2015).

Illes, J., Racine, E., and Kirschen, M., 'A Picture Is Worth a 1000 Words, but Which 1000?' in J. Illes (ed.) *Neuroethics* (Oxford: Oxford University Press, 2006).

Jackson, Frank 'Epiphenomenal Qualia' (1982) *Philosophical Quarterly* 32 (127) 127–36.

Jackson, F., and Pettit, P. 'In Defense of Folk Psychology' (1990) *Philosophical Studies* 59, 31–54.

Jezzard, P., Matthews, P., and Smith, S. *Functional MRI: An Introduction to Methods* (Oxford: Oxford University Press, 2003).

Jiang W. et al., 'Decoding the Processing of Lying Using Functional Connectivity MRI' (2015) *Behavioral and Brain Functions* 11, 1.

'A Functional MRI Study of Deception among Offenders with Antisocial Personality Disorder' (2013) *Neuroscience* 244, 90–8.

Jones, O., Buckholtz, J., Schall, J., and Marois, R. 'Brain-Imaging for Legal Thinkers: A Guide for the Perplexed' (2009) *Stanford Technology Law Review* 5 – available at: https://journals.law.stanford.edu/sites/default/files/stanford-technology-law-review/online/jones-brain-imaging.pdf (accessed 19/10/2016).

Kay, K. N. et al., 'Identifying Natural Images from Human Brain Activity (2008) *Nature* 452(7185), 352–5.

Keil, Geert 'Naturalism' in Dermot Moran (ed.) *The Routledge Companion to Twentieth-Century Philosophy* (London: Routledge, 2008), pp. 254–307.

Kennett, J. *Agency and Responsibility* (Oxford: Oxford University Press, 2001).

Kenny, A. *Freewill and Responsibility* (London: Routledge, 1978).

'Language and the Mind' repr. in *The Legacy of Wittgenstein* (Oxford: Blackwell, 1984).

Kiehl, K., and Hoffman, M. B. 'The Criminal Psychopath: History, Neuroscience, Treatment and Economics' (2011) *Jurimetrics* 51, 355.

Kim, J. 'The Myth of Nonreductive Materialism' (1989) *Proceedings and Addresses of the American Philosophical Association* 63(3), 31–47.

'Events as Property Exemplification' in *Supervenience and Mind: Selected Philosophical Essays* (Cambridge: Cambridge University Press, 1993).

Physicalism, or Something Near Enough (Princeton, NJ: Princeton University Press, 2005a).

'The Myth of Nonreductive Materialism' in P. K. Moser and J. D. Trout (eds.) *Contemporary Materialism* (New York: Routledge, 2005b), pp. 134–49.

'The Very Idea of Token Physicalism' in Simone Gozzano and Christopher Hill (eds.) *New Perspectives on Type-Identity: The Mental and the Physical* (Cambridge: Cambridge University Press, 2012), pp. 167–85.

Klein, G. A. 'A Recognition-Primed Decision (RPD) Model of Rapid Decision Making' in G. A. Klein, J. Orasanu, R. Calderwood, and C. E. Zsambok (eds.) *Decision Making in Action* (Westport, CT: Ablex Publishing, 1993), pp. 138–47.

Klingberg, T. 'Training and Plasticity of Working Memory' (2010) *Trends in Cognitive Sciences* 14(7), 317–24.

Knill, D. C., and Pouget, A. 'The Bayesian Brain: The Role of Uncertainty in Neural Coding and Computation' (2004) *Trends in Neurosciences* 27, 712–719.

Kochanska, G., and Aksan, N. 'Children's Conscience and Self-Regulation' (2006) *Journal of Personality* 74(6), 1587–618.

Kolb, B., Wishaw, I., and Campbell Tesky, G. *An Introduction to Brain and Behavior*, 5th ed. (New York: Worth Publishers, 2016).

Kolber, Adam J. 'Will There Be a Neurolaw Revolution?' (2014) *Indiana Law Journal* 85, 807–845.

'Free Will as a Matter of Law' in Dennis Patterson and Michael S. Pardo (eds.) *Philosophical Foundations of Law and Neuroscience* (Oxford: Oxford University Press, 2016).

Kosslyn, S., DiGirolamo, G., Thompson W., and Alpert, N. 'Mental Rotation of Objects versus Hands: Neural Mechanisms Revealed by Positron Emission Tomography' (1998) *Psychophysiology* 35, 151–161.

Kozel F. Andrew, Johnson, K. A., et al. 'Functional MRI Detection of Deception after Committing a Mock Sabotage Crime' (2009) *Journal of Forensic Science* 54, 220.

Kozel F. Andrew, Laken, S. J., et al. 'Replication of Functional MRI Detection of Deception' (2009) *Open Forensic Science Journal* 2, 6.

Kozel F. Andrew et al. 'Detecting Deception Using Functional Magnetic Resonance Imagining' (2005) *Biological Psychiatry* 58, 605.

Kravitz, D. J., Saleem, K. S., Baker, C. I., Ungerleider, L. G., and Mishkin, M. 'The Ventral Visual Pathway: An Expanded Neural Framework for the Processing of Object Quality' (2013) *Trends in Cognitive Sciences* 17(1), 26–49.

Kripke, S. *Naming and Necessity* (Cambridge, MA: Harvard University Press, 1980).

Wittgenstein on Rules and Private Language (Cambridge, MA: Harvard University Press, 1982).

Lamme, V. A. F. 'What Introspection Has to Offer, and Where Its Limits Lie' (2010) *Cognitive Neuroscience* 1(3), 232–40.

Law Commission's Report 'Expert Evidence in Criminal Proceedings in England and Wales' (2011) Law Com No 325.

Lee D. N., and Reddish, P. 'Plummeting Gannets: A Paradigm of Ecological Optics' (1981) *Nature* 293, 293–94.

Leiter, Brian, and Miller, Alexander 'Closet Dualism and Mental Causation,' (1998) *Canadian Journal of Philosophy* 28, 161–81.

Levy, Neil *Consciousness and Moral Responsibility* (Oxford: Oxford University Press, 2014a).

'Psychopaths and Blame: The Argument from Content' (2014b) *Philosophical Psychology* 27, 351.

Lewis, David, 'Reduction of Mind' in Samuel Guttenplan (ed.) *A Companion to Philosophy of Mind* (Oxford: Blackwell Publishers, 1994).

Libet, B., Gleason, C. A., Wright E. W., and Pearl, D. K. 'Time of Conscious Intention to Act in Relation to Onset of Cerebral Activity (readiness-potential): The Unconscious Initiation of a Freely Voluntary Act' (1983) *Brain* 106, 623–42.

Logothetis, N. 'What We Can and Cannot Do with fMRI' (2008) *Nature* 453, 869–78.

Lopez, R., Poy, R., Patrick, C., and Molto, J. 'Deficient Fear Conditioning and Self-Reported Psychopathy: The Role of Fearless Dominance' (2013) *Psychophysiology* 50, 210–18.

Lord Chief Justice, 'The Kalisher Lecture', 14 October 2014

Lösel, F. 'The Efficacy of Correctional Treatment: A Review and Synthesis of Meta-Evaluations' in J. McGuire (ed.) *What Works: Reducing Reoffending* (Chichester: Wiley, 1995).

Lucas, J., Raynor, P., and Vanstone, M. *Straight Thinking on Probation One Year On* (Bridgend: Mid Glamorgan Probation Service, 1992).

MacIntyre, Alasdair *Whose Justice, Which Rationality?* (South Bend, IN: University of Notre Dame Press, 1989).

Mackenzie, R., and Watts, J. 'Does Lack of a Moral Feeling Brain Indicate Moral Disability? Children Diagnosed with Callous Unemotional Traits, Emotion Regulation and the Potential of Treatment with Oxytocin' (2012) *Tizard Learning Disability Review* 17(4), 184–93.

Maguire, M., and Raynor, P. 'How the Resettlement of Prisoners Promotes Desistance from Crime: Or Does It?' (2006) *Criminology and Criminal justice* 6(1), 19–38.

Maibom, H. L. 'The Mad, the Bad, and the Psychopath' (2008) *Neuroethics*, 1(3), 167–84.

'To Treat a Psychopath' (2014) 35 *Theoretical Medicine and Bioethics* 31.

Mair, G. (ed.) *What Matters in Probation* (Cullompton: Willan, 2004).

Marcus, Eric *Rational Causation* (Cambridge, MA: Harvard University Press, 2012).

Maruna, S. *Making Good* (Washington, DC: American Psychological Association, 2001).

Maryland v. Smith, Case No. 106589C (Maryland Cir. Ct. 201).

McCabe D. P., and Castel, A. D. 'Seeing Is Believing: The Effect of Brain Images on Judgments of Scientific Reasoning' (2008) *Cognition* 107, 343–52.

McDonald, R., Dodson, M. C., Rosenfield, D., and Jouriles, E. N. 'Effects of a Parenting Intervention on Features of Psychopathy in Children (2011) *Journal of Abnormal Child Psychology* 39(7), 1013–23.

McGinn, C. 'All Machine and No Ghost?' (2012) *New Statesman*.

'What Can Your Neurons Tell You?' (July 11, 2013) *New York Review of Books*, at www.nybooks.com/articles/2013/07/11/what-can-your-neurons-tell-you/ (reviewing Jean-Pierre Changeux, *The Good, the True and the Beautiful: A Neuronal Approach* [2012]).

McIlwain, D. 'Living Strangely in Time: Emotions, Masks and Morals in Psychopathically-Inclined People' (2010) *European Journal of Analytic Philosophy* 6(1), 75.

McIvor, G., and McNeill, F. (eds.) *Beyond the Risk Paradigm in Criminal Justice* (Basingstoke: Palgrave, 2016).

McMillan, J. 'The Kindest Cut? Surgical Castration, Sex Offenders and Coercive Offers' (2013) *Journal of Medical Ethics* 40(9), 583–90.

Mecacci, G., and Haselager, P. 'Identifying Criteria for the Evaluation of the Implications of Brain Reading for Mental Privacy' (2017) *Science and Engineering Ethics* 15.

Meegan, D. 'Neuroimaging Techniques for Memory Detection' (2008) *American Journal of Bioethics* 8, 9–20.

Meijer, E., Ben-Shakhar, G., Verschuere, B., and Donchin, E. 'A Comment on Farwell (2012): Brain Fingerprinting: A Comprehensive Tutorial Review of Detection of Concealed Information with Event-related Brain Potentials' (2013) *Cognitive Neurodynamics* 7(2), 155–8.

Meixner, J. 'Liar, Liar, Jury's the Trier' (2012) *Northwestern Law Review* 106(3), 1451–88.

'Applications of Neuroscience in Criminal Law: Legal and Methodological Issues' (2015) *Current Neurology and Neuroscience Reports* 15(2), 513.

Meixner, J., and Rosenfeld, J. P. 'Countermeasure Mechanisms in a P300-Based Concealed Information Test' (2010) *Psychophysiology* 47(1), 57–65.

'Detecting Knowledge of Incidentally Acquired, Real-World Memories Using a P300-Based Concealed-Information Test' (2015) *Psychological Science* 25(11), 1994–2005.

Mele, A. *Springs of Action* (New York: Oxford University Press, 1992).

Menninger, K. *The Crime of Punishment* (New York: Viking Press, 1968).

Menninger, K. et al. *The Vital Balance: Life Process in Mental Health and Illnesses* (New York: Viking Boks, 1963).

Mertens, R., and Allen, J. J. B. 'The Role of Psychophysiology in Forensic Assessments: Deception Detection, ERPs and Virtual Reality Mock Crime Scenarios' (2008) *Psychophysiology* 45, 286–98.

M'Naghten's case (1843) 10 Cl & F 200.

Moore, M. 'Renewed Questions about the Causal Theory of Action' in Peter Geach (ed.) *Mental Acts* (London: Routledge and Kegan-Paul, 1959).

Law and Psychiatry: Rethinking the Relationship (Cambridge: Cambridge University Press, 1984), pp. 224–28.

'Mind, Brain, and the Unconscious' in P. Clarke and C. Wright (eds.) *Mind, Psychoanalysis, and Science* (Oxford: Basil Blackwell, 1988).

Act and Crime: The Implications of the Philosophy of Action for Criminal Law (Oxford: Oxford University Press, 1993, paperback edit. 2010).

Placing Blame: A General Theory of the Criminal Law (Oxford: Oxford University Press, 1997).

'Moral Reality Revisited' in *Objectivity in Ethics and Law* (Aldershot, UK: Ashgate, 2004), pp. 193–201.

Causation and Responsibility: An Essay in Law, Moral, and Metaphysics (Oxford: Oxford University Press, 2009).

'Libet's Challenge(s) to Responsible Agency' in Walter Sinnott-Armstrong and Lynn Nadel (eds.) *Conscious Will and Responsibility: A Tribute to Benjamin Libet* (Oxford: Oxford University Press, 2011), pp. 207–34.

'Responsible Choices, Desert-Based Legal Institutions, and the Challenge of Contemporary Neuroscience' (2012a) *Social Philosophy and Policy* 29, 233–79.

'The Specialness of the General Part of the Criminal Law' in Dennis Baker (ed.) *The Sanctity of Life and the Criminal Law: Essays in Honour of Glanville Williams* (Cambridge: Cambridge University Press, 2012b), pp. 69–105.

'Compatibilism(s) for Neuroscientists' in Enrique Villanueva (ed.) *Law and the Philosophy of Action* (The Netherlands: Rodopi Philosophical Studies, 2014), pp. 1–59.

'Stephen Morse and the Fundamental Psycho-Legal Error' (2016a) *Criminal Law and Philosophy* 10, 45–89.

'The Neuroscience of Volitional Excuse' in D. Patterson and M. Pardo (eds.) *Philosophical Foundations of Law and Neuroscience* (Oxford: Oxford University Press, 2016b), pp. 179–230.

Moran, R., 'Anscombe on "Practical Knowledge"' in J. Hyman and H. Steward (eds.) *Agency and Action* (Cambridge: Cambridge University Press, 2004).

Morse, Stephen J. 'Inevitable Mens Rea' (2003) Harvard Journal of Law & Public Policy 27, 51–64.

'Criminal Responsibility and the Disappearing Person' (2007) *Cardozo Law Review* 28, 2545–75.

'Determinism and the Death of Folk Psychology: Two Challenges to Responsibility from Neuroscience' (2008a) *Minnesota Journal of Law, Science & Technology* 9, 1–36.

'Psychopathy and Criminal Responsibility' (2008b) *Neuroethics* 1, 205.

'Lost in Translation? An Essay on Law and Neuroscience' (2011a) *Current Legal Issues* 13, 529.

Neuroscience and the Future of Personhood and Responsibility' in Rosen, J and Wittes, B eds. *Constitution 3.0: Freedom and Technological Change* (Washington, DC: Brookings, 2011b).

'Neuroscience, Free Will, and Responsibility' in W. Glannon (ed.) *Free Will and the Brain: Neuroscientific, Philosophical, and Legal Perspectives* (Cambridge: Cambridge University Press, 2015).

'The Inevitable Mind in the Age of Neuroscience' in Dennis Patterson and Michael S. Pardo (eds.) *Philosophical Foundations of Law and Neuroscience* (Oxford: Oxford University Press, 2016).

Murray, D., and Nahmias, E. 'Explaining Away Incompatibilist Intuitions' (2014) *Philosophy and Phenomenological Research* 88(2), 434–67.

Nachey, P., Wydell, H., O'Neil, K., Husain, M., and Kemard, C. 'The Role of the Pre-Supplementary Motor Area in the Control of Actions' (2007) *Neuroimage* 36, T155–T163.

Nadelhoffer, T., and Sinnott-Armstrong, W. P. 'Is Psychopathy a Mental Disease?' in N. Vincent (ed.) *Neuroscience and Legal Responsibility* (New York: Oxford University Press, 2013), pp. 229–55.

Nagel, E. *The Structure of Science: Problems in the Logic of Scientific Explanation* (New York: Harcourt, Brace, and World, 1961).

Nagel, T. 'What Is It Like to Be a Bat?' (1974) *Philosophical Review*, 83, 435–450.

Nahmias, E., and Murray, D. 'Experimental Philosophy on Free Will: An Error Theory for Incompatibilist Intuitions' in J. Aguilar, A. Buckareff, and K. Frankish (eds.) *New Waves in Philosophy of Action* (Basingstoke: Palgrave-Macmillian, 2010), pp. 189–215.

National Academies of the Sciences Report *The Polygraph and Lie Detection* (Washington, DC: National Academies Press, 2003).

Nelkin, D. K. 'Psychopaths, Incorrigible Racists, and the Faces of Responsibility' (2015) *Ethics* 125(2), 357–90.

Newman, J. et al., 'Attention Moderates the Fearlessness of Psychopathic Offenders' (2010) *Biological Psychiatry* 67, 66.

Ney, A. 'Defining Physicalism' (2008) *Philosophy Compass* 3(5), 1033–48.

'Microphysical Causation and the Case for Physicalism' (2016) *Analytic Philosophy* 57, 141–64.

Nichols, S., and Knobe, J. 'Moral Responsibility and Determinism: The Cognitive Science of Folk Intuitions' (2007) *Nous* 41(4), 663–85.

Nichols, S., and Roskies, A. 'Bringing Moral Responsibility Down to Earth' (2008) *Journal of Philosophy* 105(7), 371–88.

Ofen, Noa et al., 'Neural Correlates of Deception: Lying about Past Events and Personal Beliefs' (2016) *Social Cognitive and Affective Neuroscience* 12(1), 116–27.

Oppenheim, Paul, and Putnam, Hilary 'The Unity of Science as a Working Hypothesis' in H. Feigl et al. (eds.) *Minnesota Studies in the Philosophy of Science*, vol. 2, (Minneapolis: Minnesota University Press, 1958).

O'Shaughnessy, Brian *The Will* (Cambridge: Cambridge University Press, 2008).

Papineau, David *Philosophical Naturalism* (Oxford: Blackwell, 1993).

Thinking about Consciousness (Oxford: Oxford University Press, 2002).

'Causation Is Macroscopic but Not Irreducible' in S. Gibb, E. J. Lowe, and D. Ingthorsson, (eds.) *Mental Causation and Ontology* (Oxford: Oxford University Press, 2013).

'Naturalism' in Edward N. Zalta (ed.) *Stanford Encyclopedia of Philosophy*. Available at: https://plato.stanford.edu/entries/naturalism/, last accessed, September 9, 2014.

Pardo, Michael S., and Patterson, Dennis *Minds, Brains, and Law* (Oxford: Oxford University Press, 2013).

'Morse, Mind, and Mental Causation' (2014) *Criminal Law and Philosophy* 11(1), 111–26.

Patterson, Dennis 'Rethinking Duress' (2016) *Jurisprudence* 7(3), 672–7.

Pereboom, D. *Living without Free Will* (New York: Cambridge University Press, 2001).

'Free Will Skepticism and Criminal Punishment' in T. Nadelhoffer (ed.) *The Future of Punishment* (New York: Oxford University Press, 2013), pp. 49–78.

Pereboom, D., and Caruso G. 'Hard-Incompatibilist Existentialism: Neuroscience, Punishment, and Meaning in Life' in Gregg D. Caruso and Owen Flanagan (eds.) *Neuroexistentialism: Meaning, Morals, and Purpose in the Age of Neuroscience* (Oxford: Oxford University Press, 2018).

Peth, J. et al. Memory Detection Using fMRI – Does the Encoding Context Matter?' (2015) *NeuroImage* 113, 164–74.

Plassmann, H. et al. 'Marketing Actions Can Modulate Neural Representations of Experienced Pleasantness' (2008) *Proceedings of the National Academy of Sciences of the United States of America* 105(3), 1050–4.

Pockett, Susan 'The Neuroscience of Movement' in S. Pockett, W. P. Banks, and S. Gallagher (eds.) *Does Consciousness Cause Behavior?* (Cambridge, MA: MIT Press, 2006).

Poeppl, T. B. et al. 'Connectivity and Functional Profiling of Abnormal Brain Structures in Pedophilia' (2015) *Human Brain Mapping* 36(16), 2374–86.

Prichard, H. A. *Moral Obligation* (Oxford: Oxford University Press, 1949).

Putnam, Hilary 'Psychological Predicates' in W. Caplan and D. Merrill (eds.) *Art, Mind, and Religion* (Pittsburgh: University of Pittsburgh Press, 1967).

Reason, Truth and History (Cambridge: Cambridge University Press, 1981).

'The Meaning of "Meaning"' in *Mind, Language, and Reality* (Cambridge: Cambridge University Press, 1985).

Representation and Reality (Cambridge, MA: MIT Press, 1988).

R. v Cannings [2004] 2 Cr App R 63.

R. v Gokal [1999] EWCA Crim 669.

R. v Hamilton [2014] EWCA Crim 1555.

R. v Reed and Reed, [2009] EWCA Crim 2698.

R. v Weller [2010] EWCA Crim 1085.

Rakoff, J. 'Why Innocent People Plead Guilty' (2014) *New York Review of Books*.

Raynor, P. 'Is There Any Sense in Social Inquiry Reports?' (1980) *Probation Journal* 27(3), 78–84.

'Risk and Need Assessment in British Probation: The Contribution of LSI-R' (2007) *Psychology, Crime and Law* 13(2), 125–38.

'Consent to Probation in England and Wales: How It Was Abolished, and Why It Matters' (2014) *European Journal of Probation* 6(3), 296–307.

'Three Narratives of Risk: Corrections, Critique and Context' in C. Trotter et al. (eds.) *Beyond the Risk Paradigm in Criminal Justice* (Basingstoke: Palgrave, 2016).

Raynor, P., and Robinson, G. *Rehabilitation, Crime and Justice* (Basingstoke: Palgrave Macmillan, 2005).

Reesu G. V., and Brown, N. L. 'Inconsistency in Opinions of Forensic Odontologists When Considering Bite Mark Evidence' (2016) *Forensic Science International* 266, 263–70.

Reif, A., Rösler, M., Freitag, C. M., et al. 'Nature and Nurture Predispose to Violent Behavior: Serotonergic Genes and Adverse Childhood Environment' (2007) *Neuropsychopharmacology* 32(11), 2375–83.

Riemann, R., Angerleitner A., and Strelau, J. 'Genetic and Environmental Influences on Personality: A Study of Twins Reared Together Using the Self and Peer Report NEO-FFI Scales' (1997) *Journal of Personality* 65, 449–75.

Risinger, D. Michael 'Navigating Expert Reliability: Are Criminal Standards of Certainty Being Left on the Dock?' (2000) *Albany Law Review* 64(1), 99–149.

Ronson, J. *The Psychopath Test: A Journey through the Madness Industry* (New York: Riverhead Books, 2012).

Rorty, Richard *Philosophy and the Mirror of Nature* (Princeton University Press, 1979)

Rosen, Gideon 'Metaphysical Dependence: Grounding and Reduction' in B. Hale and A. Hoffmann (eds.) *Modality, Metaphysics, Logic, and Epistemology* (Oxford: Oxford University Press, 2010), pp. 109–35.

Rosenfeld, J. P. 'Brain Fingerprinting: A Critical Analysis' (2005) *Scientific Review of Mental Health Practice* 4, 20–37.

'P300 in Detecting Concealed Information' in B. Verschuere, G. Ben-Shakhar, and E. Meijer (eds.) *Memory Detection: Theory and Application of the Concealed Information Test* (Cambridge: Cambridge University Press, 2011).

Rosenfeld, J. P., Hu, X., Labkovsky, E., Meixner, J., and Winograd, M. 'Review of Recent Studies and Issues Regarding the P300-Based Complex Trial Protocol for Detection of Concealed Information' (2013) *International Journal of Psychophysiology* 90(2), 118–34.

Rosenfeld, J. P., Hu, X., and Pederson, K. 'Deception Awareness Improves P300-Based Deception Detection in Concealed Information Tests' (2012) *International Journal of Psychophysiology* 86, 114–21.

Rosenfeld, J. P., Labkovsky, E., Winograd, M., et al. 'The Complex Trial Protocol (CTP): A New, Countermeasure-Resistant, Accurate, P300-Based Method for the Detection of Concealed Information' (2008) *Psychophysiology* 45(6), 906–19.

Rosenfeld, J. P., Soskins, M., Bosh, G., and Ryan, A. 'Simple, Effective Countermeasures to P300-Based Tests of Detection of Concealed Information' (2004) *Psychophysiology* 41(2), 205–9.

Roskies, Adina L. 'Don't Panic: Self Authorship without Obscure Metaphysics' (2012) *Philosophical Perspectives* 26, 323–42.

'Brain Imaging Techniques' in Stephen J. Morse and Adina L. Roskies (eds.) *A Primer on Criminal Law and Neuroscience* (New York: Oxford University Press, 2013), pp. 37–71.

'Decision-Making and Self-Governing Systems' (2016) *Neuroethics* 1–13.

Roth, M., 'CMU Knows What's on Your Mind' (2009) *Pittsburgh Post-Gazette.* Available at: www.post-gazette.com/science/2009/01/04/CMU-knows-what-s-on-your-mind/stories/200901040263.

Rouse, Joseph 'Social Practices and Normativity' (2007) *Philosophy of the Social Sciences* 37(1), 49–53.

Royal College of Paediatrics and Child Health 'Sudden Unexpected Death in Infancy and Childhood: Multi-Agency Guidelines for Care and Investigation' Chair: Baroness Helena Kennedy QC.

Royal Society, The (2011) *Neuroscience and the Law.*

Royal Statistical Society (2010) Practitioner Guide; 'Fundamentals of Probability and Statistical Evidence in Criminal Proceedings – Guidance for Judges, Lawyers, Forensic Scientists and Expert Witnesses.'

Rummens, S., and Cuypers, S. 'Determinism and the Paradox of Predictability' (2010) *Erkenn* 72, 233–49.

Ryle, Gilbert *The Concept of Mind* (London: Hutcheson, 1949).

Sample, I., 'Mind-reading Program Translates Brain Activity into Words' (2012) *The Guardian* Available at: www.theguardian.com/science/2012/jan/31/mind-reading-program-brain-words.

Sayre-McCord, Geoffrey 'The Many Moral Realisms' in G. Sayre-McCord (ed.) *Essays on Moral Realism* (Ithaca, NY: Cornell University Press, 1988).

Schaffer, Jonathan 'On What Grounds What' in D. Manley, D. Chalmers, and R. Wassermann (eds.) *Metametaphysics: New Essays on the Foundations of Ontology* (Oxford: Oxford University Press, 2009), pp. 347–83.

Schaffner, Kenneth 'Approaches to Reduction' (1967) *Philosophy of Science*, 34, 137–47.

Schatzki, Theodore R. *Social Practices: A Wittgensteinian Approach to Human Activity and the Social* (Cambridge: Cambridge University Press, 1996).

Schauer, Frederick 'Can Bad Science Be Good Evidence? Neuroscience, Lie Detection and Beyond' (2010) *Cornell Law Review* 95, 1191–219.

'Lie-Detection, Neuroscience, and the Law of Evidence' in Dennis Patterson and Michael S. Pardo (eds.) *Philosophical Foundations of Law and Neuroscience* (Oxford: Oxford University Press, 2016),

Schoenmakers, S. et al. 'Linear Reconstruction of Perceived Images from Human Brain Activity' (2013) *NeuroImage* 83, 951–61.

Schwenklerk, John 'Understanding "Practical Knowledge"' (2015) *Philosophers Imprint* 15(15), 1–32.

Searle, J. 'Minds, Brains, and Programs' (1980a) *Behavioral and Brain Sciences* 3, 417–24.

'The Intentionality of Intention and Action' (1980b) *Cognitive Science* 4, 47–70.

Intentionality: An Essay in the Philosophy of Mind (Cambridge: Cambridge University Press, 1983), p. 263.

'The Chinese Room' in R. A. Wilson and F. Keil (eds.) *The MIT Encyclopedia of the Cognitive Sciences* (Cambridge, MA: MIT Press, 1999).

Mind: A brief introduction (New York: Oxford University Press, 2004).

'What Your Computer Can't Know' (2014) *The New York Review of Books.*

Sellars, Wilfrid *Empiricism and the Philosophy of Mind* (Cambridge, MA: Harvard University Press, 1997).

Setiya, K. 'Intention' in *Stanford Encyclopaedia of Philosophy*, https://plato.stanford.edu/entries/intention/ last accessed on 14/05/2017.

Shaw, Elizabeth 'Direct Brain Interventions and Responsibility Enhancement' (2014) *Criminal Law and Philosophy* 8(1), 1–20.

Shiffrin, Seana Valentine *Speech Matters: On Lying, Morality, and the Law* (Princeton, NJ: Princeton University Press, 2014).

Sifferd, K. 'In Defense of the Use of Commonsense Psychology in the Criminal Law' (2006) *Law and Philosophy* 25, 571–612.

'Translating Scientific Evidence into the Language of the Folk: Executive Function as Capacity-Responsibility' in N. Vincent (ed.) *Legal Responsibility and Neuroscience* (Oxford: Oxford University Press, 2012).

'What Does It Mean to Be a Mechanism? Stephen Morse, Non-Reductivism, and Mental Causation' (2014) *Criminal Law and Philosophy* 11(1), 143–59.

Sifferd, K., Hirstein, W., and Fagan, T 'Legal Insanity and Executive Function' in *The Insanity Defense: Multidisciplinary Views on Its History, Trends, and Controversies* (Santa Barbara, CA: Praeger, 2017), pp. 215–42.

Sinnott-Armstrong, Walter *Moral Psychology: Free Will and Moral Responsibility* (Oxford: Oxford University Press, 2014).

Sip, Kamila E. et al. 'When Pinocchio's Nose Does not Grow: Belief Regarding Lie-Detectability Modulates Production of Deception' (2013) *Frontiers in Human Neuroscience* 7(1).

Skinner, B. F. *Beyond Freedom and Dignity* (New York: Knopf, 1973).

Smart, J. 'Sensations and Brain Processes' (1959) *Philosophical Review* 66, 141–56.

Smith, A. 'Attitudes, Tracing, and Control' (2015) *Journal of Applied Philosophy* 32(2), 115–32.

Smith, P. A. 'When DNA Implicates the Innocent' (2016) *Scientific American* 314(6), 11–12.

Smith, S. 'Overview of fMRI Analysis' (2004) *The British Journal of Radiology* 77, S167–S175.

Sober, E. 'The Multiple Realizability Argument against Reductionism' (1999) *Philosophy of Science* 66(4), 542–64.

Soon, C. S., Brass, M., Heinze, H. J., and Haynes, J. D. 'Unconscious Determinants of Free Decisions in the Human Brain' (2008) *Nature Neuroscience* 11, 543–45.

Soon, C. S. et al. 'Predicting Free Choices for Abstract Intentions' (2013) *Proceedings of the National Academy of Sciences* 110(15), 6217–22.

Sorensen, Roy 'Bald-Faced Lies! Lying without the Intent to Deceive' (2007) *Pacific Philosophical Quarterly* 88, 251.

Stansbury, D. E., Naselaris, T., and Gallant, J. L. 'Natural Scene Statistics Account for the Representation of Scene Categories in Human Visual Cortex' (2013) *Neuron* 79(5), 1025–34.

Strawson, Galen 'Real Naturalism' (2013) *London Review of Books* 35(18), 28–30.

Strawson, P. F. 'Freedom and Resentment' in P. F. Strawson (ed.) *Studies in the Philosophy of Thought and Action* (Oxford: Oxford University Press, 1968), pp. 71–96.

Talbert, M. 'Blame and Responsiveness to Moral Reasons: Are Psychopaths Blameworthy?' (2008) *Pacific Philosophical Quarterly* 89(4), 516–35.

Taylor, Charles 'Interpretation and the Sciences of Man' (1971) *The Review of Metaphysics* 25(1), 3–51.

Philosophical Papers: Volume 2, Philosophy and the Human Sciences (Cambridge: Cambridge University Press, 1985).

Taylor, Richard *Action and Purpose* (Englewood Cliffs, NJ: Prentice-Hall, 1965).

Teichman, Roger *The Philosophy of Elizabeth Anscombe* (Oxford: Oxford University Press, 2008).

ten Broeke, A. 'U Bent Pedofiel, Zegt de Scanner' ('You Are a Paedophile, the Scanner Says)' (2011) *Trouw*. Available at: www.trouw.nl/tr/nl/6700/Wetenschap/article/detail/3097538/2011/12/29/U-bent-pedofiel-zegt-de-scanner.dhtml.

'De Scanner Zegt het. Dus het is Echt. Punt uit' (The Scanner Says so. Hence it is Real. Period') (2014) *De Volkskrant*. www.volkskrant.nl/opinie/de-scanner-zegt-het-dus-het-is-echt-punt-uit~a3765206/.

Tew, J., and Atkinson, R. 'The Chromis Programme: From Conception to Evaluation' (2013) *Psychology, Crime & Law* 19(5), 415–31.

Thabo Meli and Others vs R [1954] 1 WLR 228 (PC).

The Who (1966) *Substitute*. Reaction Records.

Thompson, E., and Varela, F. 'Radical Embodiment: Neural Dynamics and Consciousness' (2001) *Trends in Cognitive Sciences* 5, 418–25.

Thompson, Michael *Life and Action* (Cambridge, MA: Harvard University Press, 2008).

'Anscombe's Intention and Practical Knowledge' in Anton Ford, Jennifer Hornsby, and Frederick Stoutland (eds.) *Essays on Anscombe's Intention* (Cambridge, MA: Harvard University Press, 2011).

Uithol, S. et al. 'What Do Mirror Neurons Mirror? (2011) *Philosophical Psychology* 24(5), 607–23.

Umbach, R., Berryessa, C., and Raine, A. 'Brain Imaging Research on Psychopathy: Implications for Punishment, Prediction, and Treatment in Youth and Adults' (2015) *Journal of Criminal Justice* 43(4), 295–306.

Umeå District Court (Umeå tingsrätt) (2012) 'Samurajmålet', NJA 2012.

United States v. Dunnigan, 507 U.S. 87, 94 (1993).

United States v. Semrau, 693 F.3d 510 (6th Cir. 2012.

Van Gerven, M. et al. 'The Brain-Computer Interface Cycle' (2009) *Journal of Neural Engineering* 6(4), 41001.

Van Gulick, R. 'Reduction, Emergence and Other Recent Options on the Mind/Body Problem: A Philosophic Overview' (2001) *Journal of Consciousness Studies* 8(9–10), 1–34.

Van Overwalle, F., Baetens, K., Mariën, P., and Vandekerckhove, M. 'Social Cognition and the Cerebellum: A Meta-Analysis of over 350 fMRI Studies' (2014) *NeuroImage* 86, 554–72.

Vargas, M. *Building Better Beings: A Theory of Moral Responsibility* (Oxford: Oxford University Press, 2013).

Verschuere, B., Ben-Shakhar, G., and Meijer, E. *Memory Detection: Theory and Application of the Concealed Information Test* (Cambridge: Cambridge University Press, 2011).

Vilares, I., Wesley, M., Ahn, W.-Y., et al. 'Predicting the Knowledge-Recklessness Distinction in the Human Brain' (2017) *Proceedings of the National Academy of Sciences* 114(12), 3222–7.

Vincent, N. 'Assessment and Modification of Free Will through Scientific Techniques: Two Challenges' in W Glannon (ed.) *Free Will and the Brain* (Cambridge: Cambridge University Press, 2015).

Vitiello, B., and Stoff, D. M. 'Subtypes of Aggression and Their Relevance to Child Psychiatry (1997) *Journal of the American Academy of Child & Adolescent Psychiatry* 36(3), 307–15.

Von Hirsch, A. (2010) 'Proportionality and Progressive Loss of Mitigation: Further Reflections' in J. Robertsand A. Von Hirsch (eds.) *Previous Convictions at Sentencing* (Oxford: Hart Publishing, 2010), pp. 1–16.

von Riel, Raphael, and Van Gulick, Robert 'Scientific Reduction' (2014 version) *Stanford Encyclopedia of Philosophy* 52–54.

Ward, Tony 'Expert Evidence and the Law Commission: Implementation without Legislation?' (2013) *Criminal Law Review* 7, 561–76.

Watson, G. 'Psychopathic Agency and Prudential Deficits' (2013) *Proceedings of the Aristotelian Society* 113(3), 269–92.

Weisberg, D., Keil, F., Goodstein, J., Rawson E., and Gray, J. 'The Seductive Allure of Neuroscience Explanations' (2008) *Journal of Cognitive Neuroscience* 20, 470–7.

Weisberg, D., Taylor J., and Hopkins, E. 'Deconstructing the Seductive Allure of Neuroscience Explanations' (2015) *Decision Making* 10, 429–44.

Wickens, C. 'Multiple Resources and Mental Workload' (2008) *Human Factors* 50(3), 449–55.

Williams, Meredith *Wittgenstein, Mind and Meaning: Towards a Social Conception of Mind* (London: Routledge, 1999).

Williamson, S., Hare, R. D., and Wong, S. 'Violence: Criminal Psychopaths and Their Victims (1987) *Canadian Journal of Behavioral Science* 9(4), 455–62.

Wilson v. Corestaff Services L.P., 900 N.Y.S.2d 639 (May 14, 2010).

Wilson, Adam 'The Law Commission's Recommendation on Expert Evidence: Sufficient Reliability?' (2012) 3 *Web JCLI*

Wimmer, H. and Permer, J. 'Beliefs about Beliefs: Representation and Constraining Function of Wrong Beliefs in Young Children's Understanding of Deception' (1983) *Cognition* 13, 103–28.

Winograd, M. and Rosenfeld, J. P. 'Mock Crime Application of the Complex Trial Protocol P300-based Concealed Information Test' (2011) *Psychophysiology* 48(2), 155–61.

'Countermeasure Mechanisms in the Complex Trial Protocol' (2012) *International Journal of Psychophysiology* 85(3): 305.

Wittgenstein, Ludwig *Philosophical Investigations* (Chichester: Wiley-Blackwell, 2009).

Wolpe, P. R., 'Is My Mind Mine?' (2009) *Forbes*, Available at: www.forbes.com/2009/10/09/neuroimaging-neuroscience-mind-reading-opinions-contributors-paul-root-wolpe.html.

Yaffe, Gideon 'Mind-Reading by Brain-Reading and Criminal Responsibility' in Dennis Patterson and Michael S. Pardo (eds.) *Philosophical Foundations of Law and Neuroscience* (Oxford: Oxford University Press, 2016).

Zamble, E., and Quinsey, V. *The Criminal Recidivism Process* (Cambridge: Cambridge University Press, 1997).

Zlodre, J., Yiend, J., Burns, T., and Fazel, S. 'Coercion, Competence, and Consent in Offenders with Personality Disorder' (2016) *Psychology, Crime & Law* 22(4), 315.

INDEX